About the Author

Husam A. Alshareef earned a bachelor's degree in civil engineering from Al-Mustansiriyah University (Baghdad, Iraq); a master's degree in building construction and facility management; track integrated project delivery method (IPD) from Georgia Institute of Technology (Atlanta, GA); and a PhD in civil engineering from Kansas State University (Manhattan, KS). He is a LEED-accredited professional with expertise in sustainable design and construction.

Dr. Alshareef is an assistant professor and the director of the construction management program at Colorado State University–Pueblo (CSU-Pueblo). He also works as an adjunct professor at Kansas State University. Dr. Alshareef has around 16 years of combined experience between academia and industry, which has been utilized to write this book. Dr. Alshareef has published several research papers, some of which were published in prestigious journals. He was recognized by CSU-Pueblo as an OER hero for writing the e-book *Construction Management from a Modernized Perspective*. He is a co-chair of the Award and Celebration Committee and a member of the Colorado Market Leadership Advisory Board for the U.S. Green Building Council Mountain Region.

Sustainability in Construction

Sustainability in Construction
LEED Green Associate Certification Preparation

Husam A. Alshareef, PhD, LEED AP BD+C

New York Chicago San Francisco
Athens London Madrid
Mexico City Milan New Delhi
Singapore Sydney Toronto

Library of Congress Control Number: 2022057229

ISBN 978-1-265-01281-6
MHID 1-265-01281-4

Sponsoring Editor Ania Levinson	**Proofreader** Amitha Karkera
Editorial Supervisor Janet Walden	**Indexer** Edwin Durbin
Project Manager Tasneem Kauser, KnowledgeWorks Global Ltd.	**Production Supervisor** Lynn M. Messina
Acquisitions Coordinator Elizabeth M. Houde	**Composition** KnowledgeWorks Global Ltd.
Copy Editor Kate Brown	**Illustration** KnowledgeWorks Global Ltd.
	Art Director, Cover Jeff Weeks

To my lovely wife, Safa,
who has blessed me with her love and
taught me what it means to be a caring human being.

Contents

Preface

Sustainable practices support ecological, human, and economic health and vitality. Sustainability presumes that resources are finite and should be used conservatively and wisely with a view to long-term priorities and consequences of the ways in which resources are used. Therefore, people must work together and collaborate to be able to explore sustainable ideas and approaches that create a reasonable balance among health, social equity, and economic vitality. Since people are the ones who are responsible for all the sustainability initiatives, and since people come from different backgrounds and disciplines, an effective rating system must have a clearly defined holistic plan. Thus, the Leadership in Energy and Environmental Design (LEED) rating system and its categories are utilized to provide a framework for healthy, highly efficient, and cost-saving green building. This book addresses the core concept and principles of the LEED rating system and goes beyond the LEED Green Associate level to touch on more practical aspects of LEED and sustainability. The book also uses questions and elaborated answers with discussion so issues and their answers can be understood from various perspectives. Industry professionals and educators from across disciplines can use this book, so the book can be utilized by engineers, construction managers, architects, those in business management and hospitality, government agencies, and many others whose jobs are in tandem with sustainability whether directly or indirectly.

The book contains 10 chapters that cover approaches and applications in sustainability and address LEED Green Associate in advanced and elaborated manners. For example, Chapter 1 is the Introduction. Chapter 2 takes readers into many of the core concepts of the LEED rating system and LEED impact categories. Chapter 3 provides an overview of the U.S. Green Building Council (USGBC), the structure and scope of the LEED rating system, and the appropriate rating system adoption for a project. Chapters 4 through 10 provide a comprehensive review of the LEED rating system credit categories and their intents and strategies. A mock-up exam (like the actual LEED Green Associate exam) contains 100 questions/elaborated answers. Each chapter provides a deep dive into the LEED categories and their adoption in a more extensive and elaborated approach. The book also contains appendices:

- Appendix A contains two case studies (real-world projects). The two case studies contain graphs, diagrams, photos, and tables. One of the case studies contains calculation worksheets with detailed explanations for a real-world project.

- For reference purposes, Appendix B contains LEED scorecards of the eight adoptions: Core and Shell, Data Centers, Healthcare, Hospitality, New Construction, Retail, Schools, and Warehouses and Distribution Centers.

This book is distinguished from other LEED books on the market because it discusses sustainability beyond LEED. The book not only helps people study and prepare for the LEED Green Associate exam, but also discusses in detail the LEED categories with practical and technical approaches. In addition to the textbook, which contains narratives, questions and elaborated answers, and discussions of the questions and answers, the book contains additional resources:

- Two videos contain PowerPoint slides, graphs, diagrams, explanations of the LEED, and topics equivalent to the LEED in the current market.
- A test bank contains around 230 questions and answers not used in the textbook. The text bank is divided into six sections.
- Bonus material for Chapter 1 presents the Home Energy Rating System (HERS).

These resources can be accessed at https://www.mhprofessional.com/ SustainabilityInConstrLEED.

Developing this book, I utilized my several years of teaching LEED materials, my educational background, and over 16 years of combined experience between academia and industry. Everything in the textbook and its resources is original and from my own experience. This book and its resources help people practice critical thinking and develop a sense of curiosity about more eco-friendly applications.

Sincerely,

Dr. Husam Alshareef, LEED AP BD+C
Assistant Professor
Colorado State University–Pueblo

Outline of the Book Structure and Content

This section provides an overview of the textbook's structure and content with a description of each chapter. These descriptions illustrate the topics covered and the extent of synergy beyond the topics. The book contains 10 chapters, a LEED Terminologies and Definitions section, a mock-up exam, two appendices, and a bibliography. Topical videos, a test bank, and more about rating systems are all online additional resources. These resources can be accessed at https://www.mhprofessional.com/SustainabilityInConstrLEED.

Chapter 1, Introduction: The introduction covers sustainability topics; answers the question "Why make your building green?"; discusses in detail the background on LEED; identifies and discusses other rating systems besides LEED; and explains in detail the integrative process (IP) credit.

Chapter 2, LEED Core Concepts: The chapter discusses the LEED Core concepts in depth and then addresses questions and their elaborated answers, which all focus on topics such as impact categories, life-cycle approach and assessment, life-cycle costing, triple bottom line, regenerative building, integrative process, iterative process, systems thinking, open systems, closed systems, and feedback-loop and leverage points.

Chapter 3, Overview of USGBC and LEED: The chapter provides an overview of USGBC and LEED and then discusses questions and their elaborated answers, which all focus on topics such as community involvement and the five LEED rating systems (LEED for BD+C, LEED for ID+C, LEED for O&M, LEED for Homes, LEED for ND). Additionally, the questions and their answers focus on topics such as prerequisites and credits, project registration, certification process and levels, minimum program requirements, impact categories, pilot credit library, and LEED certifications for individuals.

Chapter 4, Location and Transportation (LT): The chapter discusses the Location and Transportation category in detail and then discusses questions and their elaborated answers, which all focus on topics such as location and transportation, location strategy and intents, transportation strategy and intents, site development and strategies, and health and livability.

Chapter 5, Sustainable Sites (SS): The chapter discusses the Sustainable Sites category in depth and then discusses questions and their elaborated answers, which all focus on topics such as sustainable sites and environmental impacts, site design and management, rainwater management, heat island effect, and light pollution and its strategies.

Chapter 6, Water Efficiency (WE): The chapter discusses the Water Efficiency category in detail and then provides questions and their elaborated answers, which all focus on topics such as innovative approaches to water conservation, full-time equivalent (FTE), flush fixtures and flow fixtures, increasing water efficiency, reducing indoor water use and some effective strategies, and reducing outdoor water use and some effective strategies.

Chapter 7, Energy and Atmosphere (EA): The chapter discusses the Energy and Atmosphere in detail and then discusses questions and their elaborated answers, which all focus on topics such as reducing energy demand and some effective strategies, increasing energy efficiency and some effective strategies, producing renewable energy strategies, refrigerant management and some effective strategies, and ongoing energy performance.

Chapter 8, Materials and Resources (MR): This chapter discusses the Materials and Resources category and then discusses questions and their elaborated answers, which all focus on topics such as materials and resources strategies, product attributes and disclosure, materials conservation and effective strategies, environmentally preferable materials, and waste management and some effective strategies.

Chapter 9, Indoor Environmental Quality (IEQ): The chapter discusses the Indoor Environmental Quality category in detail and then discusses questions and their elaborated answers, which all focus on topics such as the indoor environment, increasing indoor environmental quality, indoor air quality and some effective strategies, occupant comfort, health and functional strategies, space categorization, occupied versus unoccupied spaces, regularly versus nonregularly used spaces, individual versus shared multioccupant spaces.

Chapter 10, Innovation (IN) and Regional Priority (RP): The chapter discusses the Innovation and Regional Priority categories in detail and then discusses questions and their elaborated answers, which all focus on topics such as innovation intents and strategies, regional priority intents, and some effective strategies.

LEED Terminologies and Definitions: Readers are encouraged to review and study all the technical terms to be familiar with the LEED vocabularies and standard definitions. Additionally, several of the technical terms are explained with elaboration concerning their requirements and applicability to project sites. All the terms are sorted in alphabetical order for easy reference.

Mock-up Exam: The mock-up exam contains 100 questions, 100 answers, and elaborate explanation for each question's multiple-choice answers to help the reader understand the reasoning behind each answer. The mock-up exam was designed to mimic the LEED Green Associate exam.

Appendices

- **Appendix A, Real-World Case Studies:**
  - Involving Stakeholders at Early Stage of the Design Process to Improve Credit Points Allocation
  - LEED Process Assessments and Efficiency Improvements for Renovated Buildings
- **Appendix B, LEED Scorecards:** The appendix consists of all the building adoptions needed.

Stop and Read!

Next, you will find information about the LEED Green Associate exam and about this book and its materials.

According to the USGBC, the LEED Green Associate exam assesses your abilities at three cognitive levels: recall, application, and analysis.

- Recall Questions: These questions assess your ability to recall factual material that is described in the exam reference.

- Application Questions: These questions provide you with a novel problem or scenario to solve using familiar principles or procedures described in the exam reference.

- Analysis Questions: These questions assess your ability to break the problem down into its components to create a solution. You must recognize the different elements of the problem and evaluate the relationship or interactions of these elements.

The LEED Green Associate exam contains 100 multiple-choice questions and is delivered in a 2-hour period. The exam has scored questions and unscored questions. All questions are placed randomly throughout the exam, and candidates are not informed of a question's status, so you should respond to all questions on the exam. Unscored questions are used to gather data regarding how the question performs. These data inform the use of the question on future exams. The LEED Green Associate exam is scored between 125 and 200 points. A score of 170 or higher is required to pass.

Since the LEED Green Associate exam contains 100 questions timed for only 120 minutes, logically the designated time for each question and its answer should be 1 minute and 20 seconds. However, considering the stress and anxiety that one might have during the exam, I recommend a time designation of 1 minute for each question and its answer. Thus, when practicing on the questions and answers in this book, you should time yourself for 1 minute on each question and its answer. Having said that, the timing should be done for a group of questions, NOT individual questions. Start with timing 30 questions and their answers; then, once you feel more comfortable, increase the number to 50 questions and their answers. Once you feel comfortable, increase the number of questions to 75, and then go to the full 100 questions.

When Practicing, Please Follow These Instructions

1. Read questions and their answers for each chapter. Do not take the mock-up exam until you are finished with all the chapters in this book.

2. Read and understand problem statements closely.

3. Eliminate unrelated choices and select the correct answer.

4. Whether you select the correct OR incorrect answer, read and understand the other choices. I find this approach is extremely useful for understanding the bigger picture; also, it will help you with future questions and answers.

5. Practice the questions and their answers over and over until everything starts to make more sense to you to grasp the concept fully. Remember, it is not only about getting the correct answers, but more about understanding the concept and right practice of LEED and to be a successful LEED professional.

When Taking the Mock-up Exam, Please Follow These Instructions

1. Time yourself; remember that the actual exam is timed, so you should practice from the get-go.

2. Read the problem statements carefully. Try to read and understand the problem statement at your first glance; do not dwell on reading problem statements longer, and time should be saved for multiple-answer choices. Remember, the exam is timed, and some problem statements are lengthier than others, so practice more on reading and answering questions faster. Do not sign up for the exam without mastering this skill.

3. After reading a problem statement, start reading and eliminating answers and do not read the eliminated answers again.

4. Select the answer that makes more sense to you.

How Will I Know If I Am Ready for the LEED Green Associate Exam?

The questions and their answers in this book were developed to challenge your understanding of the LEED practice and to push you to think deeper. The questions and their answers were also made to enhance a keen eye for details and to pay attention to keywords when reading problem statements for the first time. In my opinion, if you score 80 and above on each chapter's questions and the mock-up exam, you are ready for the LEED Green Associate exam.

Note: This book is unique in that it prepares you for the LEED Green Associate exam by following a different approach. The book was designed to help you understand the materials through narrative and questions and their elaborated answers. The questions and elaborated answers were created and incorporated as study materials to provide you with a comprehensive understanding of the exam. If you study just the narrative in each chapter, you will not be able to grasp all the concepts, but with the questions and answers, everything will be clearer. The glossary is an extremely helpful resource as well.

I also highly recommend watching and comprehending the accompanying online videos and their contents before diving into the book's materials. The videos provide you with an overview of sustainability and LEED projects. Also, take advantage of the test bank to ensure a good understanding of the materials and a better preparation for the exam.

Good luck!

CHAPTER 1

Introduction

Sustainability

Growing populations and affluence around the globe have put increasing pressure on natural resources, including air and water, arable land, and raw materials. Concern over the ability of natural resources and environmental systems to support the needs and wants of global populations, now and in the future, is part of an emerging awareness of the concept of sustainability. Developing new technologies that address societal needs and wants within the constraints imposed by natural resources and environmental systems is one of the most important challenges of the 21st century. Professionals who are pursuing and working on sustainability play a central role in addressing that challenge.

Sustainable practices support ecological, human, and economic health and vitality. Sustainability presumes that resources are finite and should be used conservatively and wisely with a view to long-term priorities and consequences of the ways in which resources are used. Therefore, people must work together and collaborate to be able to explore sustainable ideas and approaches that create a reasonable balance among health, social equity, and economic vitality. Since people are the ones who are responsible for all the sustainability initiatives and since people come from different backgrounds and disciplines, an effective rating system must have a place in a clearly defined holistic plan. Thus, the Leadership in Energy and Environmental Design (LEED) rating system and its categories are utilized to provide a framework for healthy, highly efficient, and cost-saving green building.

Why Make Your Building Green?

The environmental impact of the building design, construction, and operations industry is enormous. Buildings annually consume more than 30 percent of the total energy and more than 60 percent of the electricity used in the United States. In 2021, the commercial building sector produced more than 1 billion metric tons of carbon dioxide, an increase of more than 30 percent over 1990 levels (Department of Energy [DOE], 2021). Each day 5 billion gallons of potable water are used solely to flush toilets. According to the Office of the Federal Environmental Executive (December 2020), a typical North American commercial building generates about 1.6 pounds of solid waste per employee per day; in a building with 1500 employees, this can result in an amount of 300 tons of waste per year. Development alters land from natural, biologically diverse habitats to hardscape that is impervious and devoid of biodiversity. The far-reaching influence of the built environment necessitates action to reduce its impact.

Green building practices can substantially reduce or eliminate negative environmental impacts through high-performance, market-leading design, construction, and operation practices. As an added benefit, green operations and management reduce operating costs, enhance building marketability, increase work productivity, and reduce potential liability resulting from indoor air quality problems.

Background on LEED

Following the formation of the U.S. Green Building Council (USGBC) in 1993, the organization's members quickly realized that the sustainable building industry needed a system to define and measure "green buildings." USGBC began to research existing green building metrics and rating systems. Less than a year after formation, the members acted on the initial findings by establishing a committee to focus solely on this topic. The composition of the committee was diverse; it included an architect, real estate agents, a building owner, a lawyer, an environmentalist, and industry representatives. This cross-section of people and professionals added a richness and depth to both the process and the ultimate product.

The first LEED pilot project program, also referred to as LEED version 1.0, was launched at the USGBC Membership Summit in August 1998. After extensive modifications, the LEED Green Building Rating System version 2.0 was released in March 2000 with an expansion from 40 to 69 credits and four certification achievements (EPA [November, 2003]). LEED version 2.1 followed in 2002, and LEED version 2.2 followed in 2005. LEED version 3.0 was launched in 2007 with the ability to earn up to 110 points (Fowler, 2006). LEED version 3.0 was also updated in 2009 and called LEED 2009; that remained in use until LEED version 4.0 was released in December 2013. Recently, the USGBC announced that the new version of the LEED green building program LEED version 4.1 has been released for cities, communities, and homes.

As LEED has evolved and matured, the program has undertaken new initiatives. The LEED Green Building Rating Systems are voluntary, consensus-based, and market-driven. Based on existing and proven technology, they evaluate environmental performance from a whole-building perspective over a building's life cycle, providing a definitive standard for what constitutes a green building in design, construction, and operation.

Other Rating Systems or Certifications in Addition to the LEED Rating System

WELL Building Standard

The WELL Building Standard is a performance-based system for measuring, certifying, and monitoring features of the built environment that impact human health and well-being through the seven concepts shown in Fig. 1.1. This information is in accordance with the International WELL Building Institute (IWBI). The IWBI is a leading global

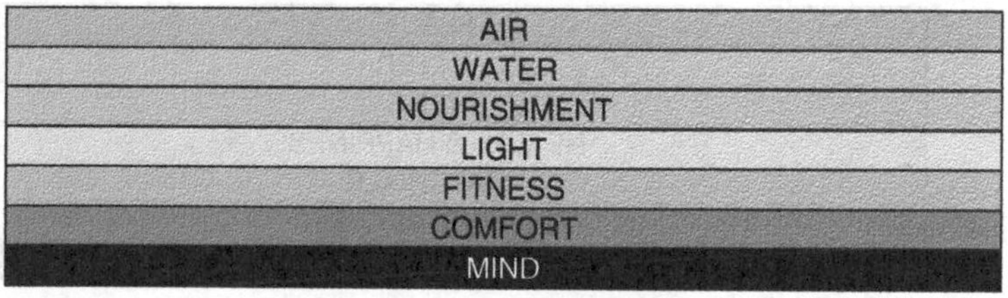

| AIR |
| WATER |
| NOURISHMENT |
| LIGHT |
| FITNESS |
| COMFORT |
| MIND |

Figure 1.1 Seven concepts of the WELL Building Standard.

movement to transform health and well-being. A public benefit corporation whose mission is to improve human health and well-being through the built environment. WELL is managed and administered by the International WELL Building Institute (IWBI).

1. **Air:** Optimize and achieve indoor air quality. Strategies include removal of airborne contaminants, prevention, and purification.
2. **Water:** Optimize water quality while promoting accessibility. Strategies include removal of contaminants through filtration and treatment and strategic placement.
3. **Nourishment:** Encourage healthy eating habits by providing occupants with healthier food choices, behavioral cues, and knowledge about nutrient quality.
4. **Light:** Minimize disruption to the body's circadian rhythm. Requirements for window performance and design, light output and lighting controls, and task-appropriate illumination levels are included to improve energy, mood, and productivity.
5. **Fitness:** Utilize building design technologies and knowledge-based strategies to encourage physical activity. Requirements are designed to provide numerous opportunities for activity and exertion, enabling occupants to accommodate fitness regimens within their daily schedule.
6. **Comfort:** Create an indoor environment that is distraction-free, productive, and soothing. Solutions include design standards and recommendations, thermal and acoustic controllability, and policy implementation covering acoustic and thermal parameters that are known sources of discomfort.
7. **Mind:** Support mental and emotional health, providing the occupant with regular feedback and knowledge about their environment through design elements, relaxation spaces, and state-of-the-art technology.

ENVISION Rating System

Envision is a flexible system of criteria and performance objectives to aid decision makers and help project teams identify sustainable approaches during the planning, design, and construction of infrastructure projects that will continue throughout the project's operations and maintenance and end-of-life phases.

| PLACE |
| WATER |
| ENERGY |
| HEALTH & HAPPINESS |
| MATERIALS |
| EQUITY |
| BEAUTY |

FIGURE 1.2   "Petals" or the seven performance categories.

The Institute for Sustainable Infrastructure (ISI) is the organization that developed and manages Envision, a framework that encourages systemic changes in the planning, design, and delivery of sustainable, resilient, and equitable civil infrastructure through education, training, and third-party project verification.

Living Building Challenge

The Living Building Challenge (LBC) is the ultimate green building standard that can be applied to any building type around the world. The goal is to create Living Buildings that incorporate regenerative design solutions that actually improve the local environment rather than simply reduce harm.

The LBC comprises seven performance categories, or "Petals," shown in Fig. 1.2. This rating system is the most optimistic system among those mentioned previously and aims to produce development with zero waste. The term "Petals" is used by the International Living Future Institute (ILFI) and the LBC community. A Petal is one of the brightly colored leaves of the corolla of a flower. The ILFI refers to its seven performance categories as "Petals." The LBC is administered by the ILFI, which its mission is "a society that is socially just, culturally rich, and ecologically restorative."

Integrative Process

An integrative process (IP) is a comprehensive approach to building systems and equipment. Project team members look for synergies among systems and components: the mutual advantages that can help achieve high levels of building performance, human comfort, and environmental benefits. The process should involve rigorous questioning and coordination and challenge typical project assumption. Team members collaborate to enhance the efficiency and effectiveness of every system.

The IP credit goes beyond checklists and encourages integration during early design stages, when clarifying the owner's aspirations, performance goals, and project needs will be most effective in improving performance. An IP comprises three phases. The first phase is called discovery, which is also the most important phase and can be seen as an expansion of what is conventionally called predesign. Actions taken during discovery are essential to achieving a project's environmental goals cost-effectively. The second phase is called the implementation phase, which combines design and construction; it begins with schematic design, is followed by design development, and ends with construction. Unlike its conventional counterpart, however, in the IP, design will incorporate all of the collective understanding of system interactions that were found during the discovery phase. The third phase is the period of occupancy, operations, and performance and feedback. Here, the integrative process measures performance and sets up feedback mechanisms. Feedback is critical to determining success in achieving performance targets, informing building operations, and taking corrective action when targets are missed.

A fully IP accounts for the interaction among all building and site systems; this credit serves as an introduction to the comprehensive process, rewarding project teams that apply an integrative approach to energy and the water system. By understanding building system interrelationships, project teams will ideally discover unique opportunities for innovative design, increased building performance, and greater environmental benefits that will earn more LEED points. By identifying synergies between systems, teams will save time and money in both the short and the long term while optimizing resource use. Finally, the IP can avoid delays and costs resulting from design changes during the construction documents phase and can reduce change orders during construction. Through the IP, a project team can more effectively use LEED as a comprehensive tool for identifying interrelated issues to develop synergistic strategies. When applied properly, the IP reveals the degree to which LEED credits are related, rather than individual items on a checklist.

To depict LEED-based projects using the IP approach, Fig. 1.3 illustrates the sequence of the phase in correlation with the timeline. Also, the figure indicates the sequence of the IP phases in relation to the conventional method phases. As can be seen from the figure, description below the timeline represents the work process during the conventional method, starting with programming and predesign. During this phase, project teams help the owner create a project program, develop owner's expectations, and create mutual goals and objectives. Once this phase is complete, the team moves to the schematic design." During this phase, the project team develops basic building parameters and an overall scheme to help the owner understand the major elements of the project. Once this phase is complete, the team moves to the design development phase. During this phase, all the project's items and details are finalized. Once this phase is complete, the team develops the construction documents to prepare the project for bidding. Once a contractor wins the bid, the contractor enters construction, which is when the physical building is completed. Afterward, the occupancy phase is the last phase, when the project users occupy and use the building.

The LEED-based project is different in terms of scope of work and process sequence. The LEED-based project consists of three phases: discovery phase; implementation phase; and occupancy, operation, and performance and feedback phase. The discovery phase, which is responsible for collecting and analyzing information, is the equivalent

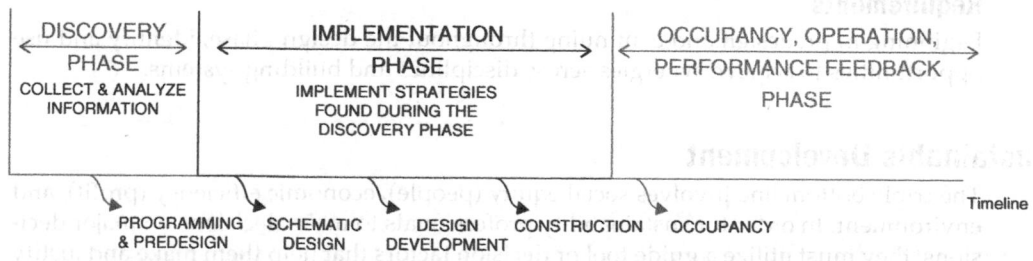

FIGURE 1.3   Integrative process phases in a sequence timeline compared with conventional method phases.

Building Adoption	Designed Points
New construction	1
Core & shell	1
Schools	1
Retail	1
Data centers	1
Warehouses and distribution centers	1
Hospitality	1–5
Healthcare	1–5

TABLE **1.1**   Building Adoptions and Their Designated Points

of programming and predesigning in the conventional method. During this phase, the project team collects and analyzes project data and information to help the owner develop the project program and other aspects of the project (equivalent to the programming and predesign phase).

The IP implementation phase, which is responsible for implementing strategies founded during the discovery phase, is equivalent to schematic design, design development, and construction in the conventional method. The third phase (occupancy, operation, and performance feedback phase) is responsible for operation, maintenance, and building reuse, repurpose, or demolition, which is equivalent to the occupancy phase in the conventional method. This phase puts some responsibilities on the building users and the owner to report back to the project team concerning design deficiencies if there are any, the construction means and methods, and other aspects of the project. This phase helps the project team eliminate mistakes and become more productive and efficient in future projects.

The IP credit applies to only the building adoptions shown in Table 1.1. This means the adoptions from the table are allowed to use the IP credit with limited reward credit points depending on the adoption type.

Intent of the Credit

To support high-performance, cost-effective project outcomes through an early analysis of the interrelationships among systems.

Requirements

Beginning in predesign and continuing throughout the design phase, identify and use opportunities to achieve synergies across disciplines and building systems.

Sustainable Development

The triple bottom line involves social equity (people), economic efficiency (profit), and environment. In order for sustainability professionals to make decisions on major decisions, they must utilize a guide tool or decision factors that help them make and justify decisions. Thus, the triple bottom line is the best tool currently to fit these criteria. The triple bottom line addresses people in a social equity approach. For instance, the living

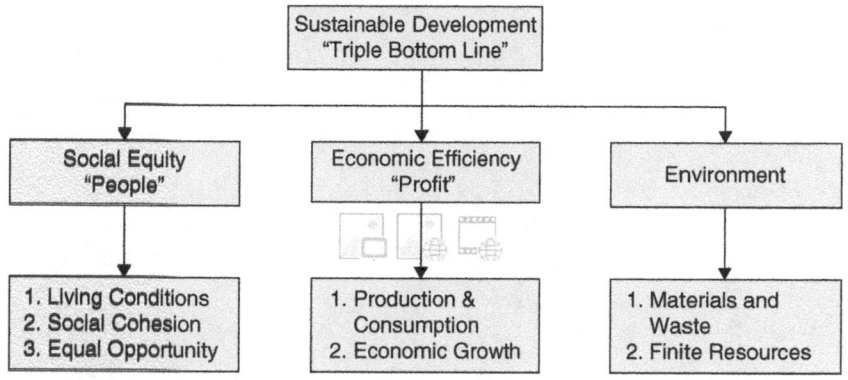

FIGURE 1.4 Sustainable development in the form of the triple bottom line.

condition of people is analyzed and assessed to ensure people are satisfied when using the building and result in increased productivity. Social cohesion is also a decision factor to ensure the sense of community is provided. Lastly, equal opportunities are provided to all who use the building and live in a community. The second decision factor is profit in the form of economic efficiency. It addresses the production of building elements and methods of consumption. The life cycle supports economic growth and prosperity. The environment decision factor addresses the utilization of raw materials from extraction, utilization, and waste. It concerns the handling of waste without destroying the environment and reserving resources for future generations. It treats resources as finite and encourages developing approaches to use them sustainably. Figure 1.4 shows the structure of sustainable development utilizing the triple bottom line approach.

References

Energy Information Administration. Emission of Greenhouse Gas Report. Report of Department of Energy (DOE) 2014/EIA-0573. Data collection from 2010 through 2020, access date was 2021. https://www.eia.gov/electricity/annual/html/epa_04_01.html.

Fowler, K. M., and E. M. Rauch. Sustainable Building Rating Systems Summary. Pacific Northwest National Laboratory operated for the U.S. Department of Energy by Battelle. July 2006. https://www.pnnl.gov/main/publications/external/technical_reports/PNNL-15858.pdf.

Office of the Federal Environmental Executive (December 2020). Guiding Principles for Sustainable Federal Buildings. Access date was 2021. https://www.sustainability.gov/pdfs/guiding_principles_for_sustainable_federal_buildings.pdf.

EPA (November, 2003), "White Paper on Sustainability, a Report on the Green Building Movement." Building Design & Construction (BDC). Access date was 2021. https://archive.epa.gov/greenbu3/greenbuilding/web/pdf/bdcwhitepaperr2.pdf.

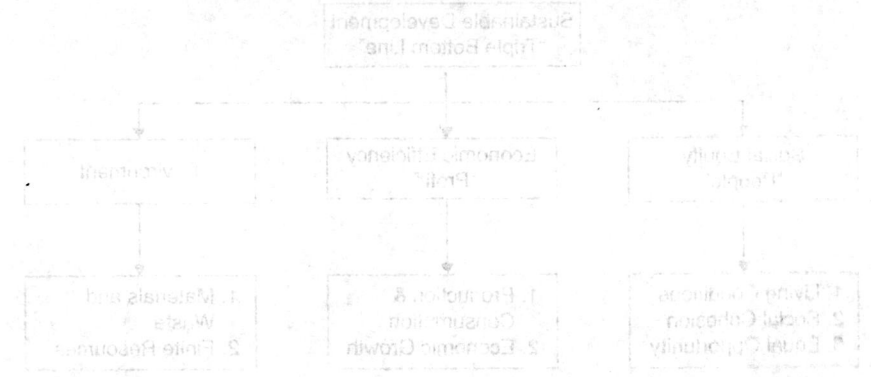

Figure 1.2   Sustainable development in the form of the triple bottom line

result ... of people is analyzed and assessed to ensure people are satisfied when using the building and result in increased productivity. So that action is also a decision factor to ensure the sense of community is provided. Lastly, peace of mind must be provided to all who use the building and live in a community. Economic is the third key. As a result in the form of economic efficiency, it addresses the production of outputs and outputs and outputs of consumption. This, in turn, supports economic growth and prosperity. The environment and economic are both addresses the utilization of raw materials extraction, utilization, and waste. It concerns the handling of waste without damaging the environment and reserving resources for future generations. It treats resources as finite and encourages developing approaches to use them sustainably.

Figure 1.2 shows the structure of sustainable development utilizing the triple bottom line approach.

References

Energy Information Administration, Inventory of Greenhouse Gas Emissions: Report of Department of Energy, DOE/ EIA/ 0573, Data collection from 2010 through 2020, access date 2021. Online: //www.eia.gov/electricity/annual/html/epa_03_01.html.

Fowler, K. M, and E. M. Rauch, Sustainable Building Rating Systems Summary, Pacific Northwest National Laboratory operated for the U.S. Department of Energy by Battelle, July 2006. https://www.pnnl.gov/main/publications/external/technical_report/PNNL-15858.pdf.

Office of the Federal Environmental Executive (December 2020), Guiding Principles for Sustainable Federal Buildings Access data 2021. https://www.sustainability.gov/pdfs/guiding_principles_for_sustainable_fed_bldgs.pdf.

EPA (November 2005), "White Paper on Sustainability, a Report on the Green Building Movement", Building Design & Construction (BDC) Access date was 2021. https://archive.epa.gov/greenbuilding/web/pdf/sciewhitepaper2.pdf.

CHAPTER 2

LEED Core Concepts

Leadership in Energy and Environmental Design (LEED) seeks to optimize the use of natural resources, promote regenerative and restorative strategies, maximize the positive and minimize the negative environmental and human health consequences of the construction industry, and provide high-quality indoor environments for building occupants. LEED emphasizes integrative design, integration of existing technology, and state-of-the-art strategies to advance expertise in green building and transform professional practice. The technical basis for LEED strikes a balance between requiring today's best practices and encouraging leadership strategies. LEED sets a challenging yet achievable set of benchmarks that define green building for interior spaces, entire structures, and whole neighborhoods.

LEED's Goals

The LEED rating systems aim to promote a transformation of the construction industry through strategies designed to achieve seven goals according to the United States Green Building Council (USGBC). These goals are listed in Fig. 2.1.

These goals are the basis for LEED's prerequisites and credits. In the BD+C rating system, the major prerequisites and credits are categorized as Location and Transportation (LT), Sustainable Sites (SS), Water Efficiency (WE), Energy and Atmosphere (EA), Materials and Resources (MR), and Indoor Environmental Quality (EQ).

The goals also drive the weighting of points toward certification. Each credit in the rating system is allocated points based on the relative importance of its contribution to the goals. The result is a weighted average: credits that most directly address the most important goals are given the greatest weight. Project teams that meet the prerequisites and earn enough credits to achieve certification have demonstrated performance that spans the goals in an integrated way. Certification is awarded at four levels (Certificated, Silver, Gold, and Platinum) to incentivize higher achievement and, in turn, faster progress toward the goals.

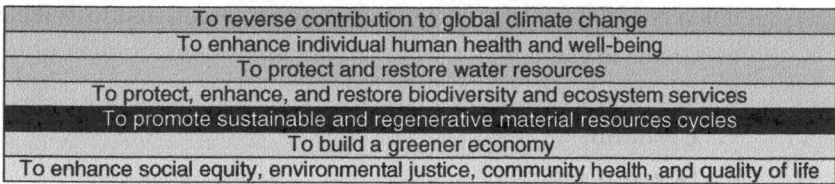

| To reverse contribution to global climate change |
| To enhance individual human health and well-being |
| To protect and restore water resources |
| To protect, enhance, and restore biodiversity and ecosystem services |
| To promote sustainable and regenerative material resources cycles |
| To build a greener economy |
| To enhance social equity, environmental justice, community health, and quality of life |

FIGURE 2.1   LEED goals according to the USGBC.

Questions

The following questions and multiple-choice answers were designed to provide a deeper insight and look into the LEED core concepts by addressing the following:

- Global climate change
- Impact categories
- Life cycle approach
- Triple bottom line
- Cost of green building
- Regenerative building
- Integrative process
- Iterative process
- System thinking

1. EA Prerequisite-Fundamental Refrigerant Management and EA Credit-Enhanced Refrigerant Management require the project to abandon which refrigerant?

   a. HCFC

   b. CFC

   c. GWP

   d. GHG

2. Human health protection throughout life cycles of projects is part of which impact category?

   a. Reverse contribution to global climate change

   b. Promote sustainable and regenerative material resources cycles

   c. Protect, enhance, and restore biodiversity and ecosystem services

   d. Enhance individual human health and well-being

3. Which impact category most directly deals with greenhouse gas emission?

   a. Promote sustainable and regenerative material resources cycles

   b. Protect, enhance, and restore biodiversity and ecosystem services

   c. Enhance social equity, environmental justice, and community quality of life

   d. Reverse contribution to global climate change

4. Activities related to the processing of materials all the way through to the delivery of the final product to the end user are referred to as what?

   a. Cradle to grave

   b. Cradle to cradle

   c. Upstream

   d. Downstream

5. What are some of the common issues a firm may encounter when dealing with owners that have never done LEED projects before? Select two.

   a. Integrative process requires extra time

   b. The time and costs associated with holding meetings during the integrative process

   c. Extended construction schedule

   d. Convincing the owner to limit their participation

6. What encourages collaboration across disciplines to involve diverse team members during the predesign phase?

   a. Regional priority

   b. Innovation

   c. System thinking

   d. Integrative process

7. An improvement in one part of a system may have a negative effect on another part of the same system. Communication across all levels in an organization should exist so that a solution in one area does not result in a problem in another area. Which of the following represents this statement?

   a. Iterative process

   b. Integrative process

   c. Feedback loop

   d. Regenerative process

   e. System thinking

8. Which of the following is the most sustainable?

   a. Neutral system

   b. Closed system

   c. Open system

   d. Leverage points

9. Triple bottom line three pillars are?

   a. Stakeholder factors

   b. Shareholder factors

   c. Social factors

   d. Environmental factors

   e. Economic factors

10. What are the three phases of the integrative process?

    a. Design and construction

    b. Occupancy, operation, and maintenance

    c. Discovery

d. Recovery and development

e. Predesign

11. What is the minimum passing score of the LEED Green Associate exam?

a. 175

b. 170

c. 70

d. 85

12. A level of efficiency for a high-performance building that produces all the energy it uses and is so energy efficient that a renewable energy system can offset all or most of its annual energy consumption. Which of the following does this statement define?

a. Aeroponics

b. Lean process improvement

c. Zero net energy

d. Vertical farm

13. Harmful organic chemical compounds that evaporate under normal indoor temperature and pressure conditions and are released by synthetic materials such as flooring, wall-covering, furniture and paint, adhesive and sealants are referred to as what?

a. Sulfur dioxide (SO$_2$)

b. Greenhouse gas emission

c. Volatile organic compounds (VOCs)

d. Xeriscaping

14. Which of the following is an example of a negative feedback loop?

a. A thermostat whose temperature feedback indicates to the system when to turn off

b. Warming oceans releasing greenhouse gases, which would warm the ocean even more

c. Increased heat islands require more mechanical cooling, which generates more greenhouse gas emission causing the need for more cooling

15. The _____ life cycle is a linear consumption to waste process.

a. Cradle to cradle

b. Cradle to grave

c. Cradle to gate

d. Downstream activity

16. According to some studies, green building costs (initial costs) are higher than traditional building by how much?

a. 2%

b. 3%

   c. 4%

   d. 5%

   e. 10%

17. A system that uses only as much water and energy as it can produce refers to what?

   a. Closed system

   b. Open system

   c. Regenerative building system

   d. Integrative process

18. USGBC has officially made the integrative process a credit in all the rating systems except for what?

   a. LEED BD+C

   b. LEED ID+C

   c. LEED ND

   d. LEED O+M

   e. LEED homes

19. At what phase are an energy modeling analysis and water budget analysis performed?

   a. During construction

   b. During feasibility study

   c. During the integrative process

   d. Discover phase

20. At which phase and in accordance with owner's project requirements, basis of design, and eventual project design should the energy modeling analysis be documented?

   a. Discovery phase

   b. Implementation phase

   c. During construction

   d. Post construction

21. A structure in which data is gathered and reported in order to meet a desired output response is referred to as what?

   a. System thinking

   b. Closed system

   c. Open system

   d. Feedback loop

22. Which of the following is an example of an automatic space heater?

   a. Closed system

   b. Open system

    c. Negative feedback loop

    d. Positive feedback loop

23. A small change can yield large results is an example of what?

    a. Positive feedback loop

    b. Leverage points

    c. Embodied energy

    d. Life cycle assessment

Answers

1. b
2. d
3. d
4. d
5. a & b
6. d
7. e
8. b
9. c, d, e
10. a, b, c
11. b
12. c
13. c
14. a
15. b   Cradle to grave is an open system where materials are extracted, manufactured, purchased, consumed, and disposed of with a clear beginning and finite end. The focus of the triple bottom line is the stakeholders.
16. a
17. c   A regenerative system is part of the closed system. It is a more specific term for using water and energy as they can be produced. Selecting option a (closed system) is incorrect because the closed system here is more general.
18. c & d
19. d
20. b
21. d
22. c   A negative feedback loop is when the system is self-corrected and stays within a particular set of parameters.
23. b

CHAPTER 3

Overview of USGBC and LEED

LEED (LEED: Leadership in Energy and Environmental Design) was developed by the United States Green Building Council (USGBC) as a framework for identifying, implementing, and measuring green building and neighborhood design, construction, operation, and maintenance. LEED is a voluntary, market-driven, consensus-based tool that serves as a guideline and assessment mechanism. LEED rating systems address commercial, institutional, and residential buildings and neighborhood developments.

LEED is designed to address environmental challenges while responding to the needs of competitive market. LEED certifications demonstrate leadership, innovation, environmental stewardship, and social responsibility. LEED gives building designers and operators the tools they need to immediately improve both building performance and the bottom line while providing healthful indoor spaces for a building's occupants.

LEED-certified buildings are designed to deliver the following benefits according to the USGBC:

- Lower operating costs and increased asset value
- Reduced waste sent to landfills
- Energy and water conservation
- More healthful and productive environments for occupants
- Reduction in greenhouse gas emissions
- Qualification for tax rebates, zoning allowances, and other incentives in many cities

By participating in LEED, owners, operators, designers, and builders make a meaningful contribution to the green building industry. By documenting and tracking building's resource use, they contribute to a growing body of knowledge that will advance research in this rapidly evolving field. This will allow future projects to build on the successes of today's designs and bring innovations to the market.

LEED Certification Process

The LEED certification process begins when the owner selects the rating system and registers the project. The project is then designed to meet the requirements for all prerequisites and for the credits the team has chosen to pursue. After documentation has been submitted for certification, a project goes through preliminary and final reviews. The preliminary review provides technical advice on credits that require additional work for achievement, and the final review contains the project's final score and certification level. The decision can be appealed if a team believes additional consideration is warranted.

LEED has four levels of certification, depending on the point thresholds achieved. This means LEED as a rating system certifies projects based on their performance and credit point allocation. Table 3.1 shows the level of certifications USBGC can offer and their reward points.

The USGBC can also certify individuals with LEED accreditation. Table 3.2 indicates the levels of certification individuals can receive and the equivalent specialty. The first certification is LEED Green Associate, which is the basic level of certification that provides general knowledge about USGBC and LEED rating system practice. Individuals who are interested in this certification can obtain it through an exam. The exam is called LEED Green Associate Exam, and it is offered by the Green Business Certification Inc (GBCI). GBCI is an American organization that provides third-party credentialing and verification for several rating systems relating to the built environment. The GBCI was created by the International Finance Corporation (IFC). A more advanced certification than the LEED Green Associate is LEED Accredited Professional (AP) with specialty. Individuals can specialize in Building Design and Construction (BD+C), Interior Design and Construction (ID+C), Operation and Maintenance (O+M), Neighborhood and Development (ND), or Homes. To obtain the LEED AP certification with specialty,

LEED Certification Level	Points Allocated
Certified	40–49
Silver	50–59
Gold	60–79
Platinum	90 points and above

TABLE 3.1 Project Certifications and Point Allocation

Certification Levels	Specialty
LEED Green Associate	
LEED AP with specialty	• Building Design and Construction (BD+C) • Interior Design and Construction (ID+C) • Maintenance and Operation (O+M) • Neighborhood and Development (ND) • Homes
LEED Fellow	

TABLE 3.2 Certification Levels of Individuals in Accordance with their Experiences and Involvements

an individual must pass the LEED Green Associate exam and remain in good standing with USGBC. Then, once that individual can be eligible to take the LEED AP with specialty exam. Once the LEED AP exam with specialty is passed, then that individual can be granted the certification. The LEED Fellow certification is the last and most prestigious certification an individual can obtain. This certification does not require examination; rather it is a nomination-based selection. An individual must have around 10 years of sustainability experience and have passed the LEED AP with specialty exam and obtain the certification. Also, individuals must have made a significant contribution to sustainability by teaching, publishing, and providing project leadership for around 8 years after passing the LEED AP with specialty exam.

Questions

The questions and multiple-choice answers for this chapter were designed to provide a deeper insight and look into LEED and the USGBC by addressing the following:

- Having a holistic understanding of the USGBC and LEED by including questions on the LEED Green Associate exam that pertain to the structure of the USGBC and LEED rating systems.
- Getting familiar with the LEED certification process.
- Understanding the LEED programs and credentials.
- Studying questions and answers that constitute the knowledge domains of the LEED process and relevant credit categories.

1. Which rating system adoption is to be used if more than 40% of the gross floor area is incomplete at the time of certification?
   a. Retail
   b. Data Center
   c. Core and Shell
   d. New Construction and Major Renovation

2. The rating system checklist contains nine categories. Which category among all the nine categories totals 33 points?
   a. Indoor Environmental Quality
   b. Material and Resources
   c. Energy and Atmosphere
   d. Water Efficiency

3. During the design and construction of a LEED project, uncertainties may arise about certain applications. To verify the applicability of these applications, what should the project team do?
   a. Find answers through Internet search engines
   b. Submit a credit interpretation request
   c. Avoid unverified applications
   d. Submit these applications anyway

4. What does LEED stand for?

   a. Leadership in Energy and Water Efficiency and Design

   b. Leadership in Energy Efficiency design

   c. Leadership in Energy and Environmental Design

   d. Leadership in Environmental Efficient Design

5. On the LEED checklist, which credit categories are found? Choose two.

   a. Indoor Air Quality

   b. LEED AP legacy

   c. Energy Efficiency

   d. Location and Transpiration

   e. Innovation

6. What determines the level of certification a project may achieve?

   a. The impact categories

   b. The integrative process and team collaboration

   c. The rating system selected

   d. The total number of points earned by the project

7. Select an additional certification besides LEED BD+C & Core and Shell a project team can earn for the same project.

   a. BD+C School

   b. LEED BD+C New Construction

   c. LEED ID+C Commercial Interior

   d. LEED BD+C Healthcare

8. What is the point of range for earning LEED Gold?

   a. 70–79

   b. 60–69

   c. 50–69

   d. 60–79

9. What is the minimum required gross floor area for a project seeking LEED BD+C and LEED O+M?

   a. There are no thresholds.

   b. It is 1000 square feet.

   c. The project team can decide the size of the project.

   d. It is 10,000 square feet.

10. What is the fee of Credit Interpretation Ruling (CIR) according to the USGBC current rating?

    a. $1200

    b. $500

    c. $220

    d. No fee

11. Under this program, organizations are offered larger discounts for certifying 25 or more projects within 3 years. Which LEED program should be pursued?

    a. Recertification program.

    b. LEED Campus program.

    c. LEED Volume program.

    d. LEED projects cannot be certified in groups.

12. Continuing education of 30 hours must be completed within 2 years of earning the credential for which certification?

    a. LEED Green Associate

    b. LEED Fellow

    c. LEED AP with specialty credentials

    d. LEED AP Indoor Environmental Quality (IEQ)

13. Which specialty credential is for professionals who are involved in planning, design, and development of neighborhoods that reduce transportation emissions, increase walkability and accessibility to open space, and incorporate green buildings and infrastructure?

    a. LEED AP BD+C

    b. LEED AP ID+C

    c. LEED AP ND

    d. LEED AP Homes

14. Which certificate provides on-site verification services for a LEED Homes project and assembles and submits the project submittals for certification review?

    a. LEED Campus

    b. Recertification program

    c. LEED Green Classroom Professional (GCP)

    d. Green Raters

15. Select the two terms that are incorrectly formatted per the USGBC trademark policy.

    a. LEED GA

    b. USGBC

    c. US Green Building Council

    d. U.S. Green Building Council

    e. LEED Green Associate

16. What is the minimum LEED level certification that can be earned?

    a. Accredited

    b. Certified

c. Approved

d. Certified with condition

17. What are some of the involvements or contributions an individual can provide to USGBC? Select three.

a. Provide community service under USGBC chapters

b. Donate money to people in need

c. Develop an educational program and materials for USGBC

d. Encourage using raw materials more often in construction projects

e. Advocate to influence legislation on green buildings and promote changes

18. A reference guide such as Building Design and Construction facilitates credentials such as LEED AP BD+C. Which adoption is used through the Building Design and Construction reference guide?

a. Healthcare

b. Existing Building

c. Multifamily Low Rise

d. Built Project

19. Under which adoption must the owner or tenant occupy at least 50% of the leasable square footage?

a. Core and Shell Development

b. Retail

c. Hospitality

d. New Construction and Major Renovation

20. Which adoption is used for an interior build out that is not complete?

a. Data Center

b. Core and Shell

c. Warehouses and Distribution Centers

d. Healthcare

21. Which rating system is applied to existing buildings that are undergoing improvements with little or no construction?

a. LEED ID+C

b. LEED Homes

c. LEED ND

d. LEED O+M

22. Which rating system is used for a project in the conceptual or master planning phase or under construction (up to 75% constructed)?

a. LEED ND

b. LEED Homes

c. LEED BD+C

d. LEED ID+C

23. The LEED categories LT, WE, and EA for the LEED Homes rating system are conducted by which choice?

    a. Performance path

    b. Prescriptive or performance paths

    c. Energy and water remolding

    d. c and d

24. Which is defined as a database of credits that are being tested for future versions of LEED?

    a. Impact category

    b. Certification level

    c. Pilot credit library

    d. Build a greener economy

    e. Rating system

25. Which individual handles all of the project submittal work through LEED online?

    a. Owner

    b. LEED professional only

    c. Project manager

    d. Project team administrator

    e. A third-party consultant

26. Using the 40/60 rule in selecting a rating system for a building, if the appropriate rating system is more than 60% of the gross floor area of the LEED project building or space in question, then what should the rating system be?

    a. It should not be selected.

    b. It should be selected.

    c. The project team must decide which rating system is most applicable.

    d. The decision ultimately is in the owner's hands.

27. If a project seeks LEED certification and is denied awarded credits during the reviewing process, how long does the project team have to accept or appeal decisions like these?

    a. 10 days

    b. 15 days

    c. 25 days

    d. 35 days

28. Select the LEED program that simplifies the certification process for buildings that do not have to have the same design or function, but must be on the same site and owned by the same owner?

    a. LEED Campus Program

    b. LEED Volume Program

    c. LEED Professional Program

    d. LEED AP

29. For an existing building, if an owner wants to continue to operate and maintain the building as set forth in the initial certification, then it is recommended to use which of the below?

    a. LEED BD+C

    b. LEED AP legacy

    c. Recertification program

    d. LEED O+M

30. What is the proper terminology for the interrelationship between credit categories, systems, and components that can be realized through the integrative process to achieve high levels of building performance, human performance, and environmental benefits?

    a. Certification level

    b. Synergy

    c. Minimum Program Requirements (MPRs)

    d. Pilot credit library

31. When a team wants to measure building performance based on LEED and building codes, what standard must the team follow?

    a. American Society of Heating, Refrigerating, and Air-Conditioning Engineering (ASHRAE)

    b. American National Standards Institute (ANSI)

    c. Illuminating Engineering Society of North American (IESNA)

    d. Local and state codes

    e. LEED standard reference or local code, whichever is more stringent

32. Which statement is true about LEED interpretation?

    a. LEED interpretation cannot be applied to more than one project.

    b. LEED interpretation can be applied to more than one project.

    c. LEED interpretation is used to add more requirements to the project.

    d. LEED interpretation is used only for the LEED BD+C rating system.

33. Which LEED category from the scorecard does not have prerequisites, only credits?

    a. Materials and Resources

    b. Energy and Atmosphere

    c. Location and Transportation

    d. Sustainable Site

34. Which LEED rating system in the list works in close association with LEED ID+C?

    a. LEED BD+C Core and Shell
    b. LEED BD+C Schools
    c. LEED BD+C Retail
    d. Hospitality

35. What is a building rating system used globally besides LEED?

    a. Green Rater
    b. Green Classroom Professional (GCP)
    c. Green Globes
    d. International Code Council (ICC)

36. What is the only thing that is used throughout the entire process of certifying a LEED building?

    a. Commissioning
    b. Commissioning agent
    c. Cooling tower water use
    d. Scorecard

37. Which other LEED certification can be earned besides LEED BD+C?

    a. LEED ID+C
    b. LEED O+M
    c. LEED Homes
    d. LEED ND

38. When can a team use the credit interpretation request?

    a. The team wants to earn extra credit.
    b. The reference guide does not address the specific issue.
    c. The team makes an appeal for a denied credit.
    d. The team wants to earn two certifications if the situation allows.

39. A four-story residential building can be certified under which rating system?

    a. Multifamily Midrise
    b. Homes and Multifamily Low Rise
    c. Healthcare
    d. Hospitality

40. What is correct about a project that earns all the credits necessary to achieve Gold LEED certification, but did not meet the prerequisite from the Energy and Atmosphere category?

    a. The project can still earn the certification.

    b. The team can request an exception and earn a temporary certification until this prerequisite is accomplished.

    c. The project cannot earn certification.

    d. It is up to the LEED AP professional on the project to decide.

41. An owner wants to develop a mixed-use building for a supermarket on the first floor, a gym on the second floor, and offices on the third and fourth floors. Which rating system should the owner pursue?

    a. LEED ID+C Commercial

    b. LEED BD+C New Construction

    c. LEED BD+C Retail

    d. LEED Homes Multifamily Low Rise

42. When there are options for two applicable rating systems, what should an architect do?

    a. Go with the most appealing one

    b. Use the 40/60 rule

    c. Go with the cheaper one

    d. Let the LEED professional on the project decide

43. Which rating system fits for the building envelope and Heating, Ventilation, and Air-Conditioning (HVAC)?

    a. LEED ID+C

    b. LEED BD+C Warehouses and Distribution Centers

    c. LEED BD+C Core and Shell

    d. LEED BD+C Retail

44. The McDonald Corporation is pursuing LEED certification for all of its branches across the globe. Which program should the corporation pursue?

    a. LEED Campus Program

    b. LEED Recertification Program

    c. Green Globes

    d. LEED Volume

45. Smart Location and Linkage is a LEED category under which of the following?

    a. Building Design and Construction

    b. Operation and Maintenance

    c. Neighborhood Development

    d. Interior Design and Construction

Answers

1. **c**

2. **c**

3. **b**

4. **c**

5. **d & e**   Options d and e are the only LEED categories, while the rest are either LEED credits or certifications.

6. **d**

7. **c**   For a project to receive Building Design and Construction (BD+C) Core and Shell certification, 40% of the building should be incomplete. Therefore, BD+C School (option a) and LEED BD+C New Construction (option b) do not apply. Option d is inapplicable because it is for healthcare.

8. **d**

9. **b**   Option b is one of the Minimum Program Requirements (MPRs).

10. **c**

11. **c**

12. **c**

13. **c**

14. **d**

15. **a & c**

16. **b**

17. **a, c, e**

18. **a**   Review closely the rating system adoptions. The Healthcare rating system is the only adoption that is certified under Building Design and Construction.

19. **a**

20. **b**

21. **d**

22. **c**

23. **b**   The project team must decide early which path must be chosen, either a performance or a prescriptive path.

24. **c**

25. **d**   During the predesign phase, a project team is assembled, and a project team administrator is assigned, who does not have to be LEED accredited.

26. **b**   If the rating system is less than 40% of the gross floor area of the LEED project, then the rating system should NOT be selected. If the rating system is more than 60%, then the rating system should be selected. If the rating system is between 40 and 60 percent, then the rating system can be decided by the project team.

27. **c**

28. **d** The LEED Volume program certifies at least 25 projects and more than that do not have the same design and do not have to be owned by the same organization.

29. **c** The recertification program is used for the LEED Maintenance and Operation (O+M) rating system for existing buildings to provide operational benefits throughout the life of the building.

30. **b** The problem statement is the definition of the word *synergy*.

31. **e** If local code is more stringent than the standard referenced in the LEED credit, then the credit requirement would be fulfilled by following the local code requirement. Conversely, if the standard referenced in the LEED credit is more stringent than the local code, then both the LEED credit and the local code will be fulfilled.

32. **b** LEED interpretation is used to clarify vague requirements or standards and can be applied to multiple projects with the same senior.

33. **c**

34. **a**

35. **c** Green Globes is a green building rating system that originated in Canada and has spread to the United States.

36. **d** The scorecard is extremely important, starting from the predesign phase through postconstruction and certification and occupancy. The project team uses the scorecard to check against whether the team is on the right track to achieve the level of certification that was agreed on.

37. **b**

38. **b**

39. **a** Multifamily residential buildings of four to eight stories above grade can be certified under Multifamily Midrise.

40. **c** All prerequisites must be achieved and then credits received so a project can earn a certification.

41. **b** Always, a mixed-use building goes with LEED Building Design and Construction (BD+C) New Construction.

42. **b** The 40/60 rule is used when there are two rating systems applicable and a team wants to decide between them.

43. **c**

44. **d** The LEED Volume Program is the best choice for certifying a large number of buildings, not necessarily on the same site.

45. **c** The Smart Location and Linkage category is the first LEED category on the scorecard of the LEED Neighborhood and Development (ND).

CHAPTER 4

Location and Transportation (LT)

The Location and Transportation (LT) category rewards thoughtful decisions about building location with credits that encourage compact development, alternative transportation, and connection with amenities such as restaurants and parks. The LT category is an outgrowth of the Sustainable Site (SS) category, which formerly covered location-related topics. Whereas the SS category now specifically addresses on-site ecosystem services, the LT category considers the existing features of the surrounding community and how this infrastructure affects the occupant's behavior and environmental performance.

Well-located buildings take advantage of existing infrastructure: public transit, street networks, pedestrian paths, bicycle networks, services and amenities, and existing utilities such as electricity, water, gas, and sewage. By recognizing existing patterns of development and land density, project teams can reduce strain on the environment from the material and ecological costs that accompany the creation of new infrastructure and hardscape. In addition, the compact communities promoted by the LT credits encourage robust and realistic alternatives to private automobile use, such as walking, biking, vehicle shares, and public transit. These incremental steps can have significant benefits: A 2009 Urban Land Institute study concluded that improvements in land use patterns and investments in public transportation infrastructure alone could reduce greenhouse gas emission from transportation in the United States by 9 to 15 percent by 2050; globally, the transportation sector is responsible for about one-quarter of energy-related greenhouse gas emission.

If integrated into the surrounding community, a building can offer distinct advantages to owners and infrastructure to the project site. For occupants, walkable and bike-able locations can enhance health by encouraging daily physical activity, and proximity to services and amenities can increase happiness and productivity. Locating in a vibrant, livable community makes the building a destination for residents, employees, customers, and visitors, and the building's occupants will contribute to the area's economic activity, creating a good model for future development. Reusing previously developed land, cleaning up brownfield sites, and investing in disadvantaged areas conserve undeveloped land and ensure efficient delivery of services and infrastructure.

Design strategies that complement the building's location are also rewarded in the LT section. For example, by limiting parking, a project can encourage building users to take alternative transportation. By providing bicycle storage, alternative fuel facilities, and preferred parking for green vehicles, a project can support users seeking transportation options.

27

Selecting and Developing the Site Wisely

Buildings affect ecosystems in a variety of ways. Development of a greenfield, or previously undeveloped site, consumes land. Development projects may also encroach on agricultural lands and wetlands or water bodies and compromise existing wildlife habitats. Choosing a previously developed site or even a damaged site that can be remediated reduces pressure on undeveloped land. Developing a master plan of the project site helps engrain environmental considerations as adaptations or expansions of site facilities occur over time. Planning for joint use of facilities integrates the project into the surrounding community and conserves material and land resources through optimized use of infrastructure.

Reducing Emissions Associated with Transportation

Environmental concerns related to buildings include vehicle emissions and the need for vehicle infrastructure as building occupants travel to and from the site. Emissions contribute to climate change, smog, acid rain, and other air quality problems. Parking areas, roadways, and building surfaces increase stormwater runoff and contribute to the urban heat island effect. In 2020, of commuters in America ages 16 and older, 76 percent drove to work alone. Of the remaining 24 percent who used alternative means of transportation (including working from home), only 5 percent used public transportation and 11 percent carpooled (U.S. Census Bureau, 2021). Locating the project near residential areas and providing occupants with cycle racks, changing facilities, preferred parking, and access to mass transit and an alternative fuel filling station can encourage use of alternative forms of transportation. Promoting mass transit reduces the energy required for transportation as well as the space needed for parking lots, which encroach on green space.

Managing Stormwater Runoff

As areas are developed and urbanized, surface permeability is reduced, which in turn increases the runoff transported via pipes and sewers to streams, rivers, lakes, bays, and oceans. Impervious surface on the site may cause stormwater runoff that harms water quality, aquatic life, and recreation opportunities in receiving waters. For instance, parking areas contribute to stormwater runoff that is contaminated with oil, fuel, lubricants, combustion by-products, material from tire wear, and deicing salts. Runoff accelerates the flow rate of waterways, increasing erosion, altering aquatic habitats, and causing erosion downstream. Effective strategies exist to control, reduce, and treat stormwater runoff before it leaves the project site.

LEED Project Boundary

For single-building developments, the LEED submittal typically covers the entire project scope and is generally limited to the site boundary. However, in some cases, a project is a portion of a larger multiple-building development. In these situations, the project team may determine the limits of the project submitted for LEED certification differently from the overall site boundaries. This LEED project boundary is the portion of the project site that is submitted for LEED certification and must be used consistently across this LEED category.

Floor Area Ratio (FAR)

It is defined as the measurement of a building's floor area in relation to the size of the lot/parcel that the building is located on. The FAR is a computation determined by dividing the total gross building floor area (square feet) by the land area of the lot. In cases where a project site encompasses several buildings on several lots, the floor area ratio may be combined and averaged over the entire project site. To calculate the maximum floor area ratio, multiply the general plan FAR by the lot square footage. The total gross floor area (square feet) of all floors of the building shall not exceed this amount.

Calculations:

General plan FAR limit = 0.4

Lot size is 20,000 square feet

$0.40 \times 20,000 = 8,000$ square feet (sqft)

The 8,000 sqft is the maximum building size

Building Density

The general plan establishes minimum and maximum densities for residential uses in all parts of the city. Residential density is a computation expressing number of dwelling units per acre based on the gross lot area prior to the dedication of any right-of-way, public parks, or other public areas. In cases where a project site encompasses more than one lot, the density may be averaged over the entire project site.

Calculations:

Lot size is 20,000 sqft/43,560 = 0.46 acres

General plan density range is 3 to 8 units/acre

3 to 8 units/acre × 0.46 acres = 1.38 to 3.68 units. This could be rounded up to 2-4 units

Keywords and Definitions

Aquifer: A body of saturated rock through which water can easily move.

Biodiversity: The variety of all life on Earth, including plants, animals, insects, microorganisms, and humans.

Erosion: A combination of processes or events by which materials of Earth's surface are loosened, dissolved, and transported by natural agents (e.g., water, wind, or gravity).

National Pollutant Discharge Elimination System (NPDES): A permit program that controls water pollution by regulating point sources that discharge pollutants into waters of the United States. Industrial, municipal, and other facilities must obtain permits if their discharges go directly to surface waters.

Sedimentation: The addition of soil particles to water bodies by natural and human-related activities. Sedimentation often decreases water quality and can accelerate the aging process of lakes, rivers, and streams.

Stormwater Runoff: Consists of water precipitation that flows over surfaces into sewer systems or receiving water bodies. All precipitation that leaves project site boundaries on the surface is considered stormwater runoff.

Stormwater Pollution Prevention Plan: Describes all measures to prevent stormwater contamination, control sedimentation and erosion during construction, and comply with the requirements of the Clean Water Act.

Questions

The questions and multiple-choice answers for this chapter were designed to provide a deeper insight and look into the LT category by addressing topics in the questions that follow.

1. Which option below is NOT the main intent of the LT category?

   a. Site development

   b. Health and livability

   c. Location

   d. Site assessment

2. Project selection is determined by

   a. The financial constraints of the owner

   b. The financial and geographical constraints of the owner

   c. The environmental effects

   d. Convenience to users

3. What is the main goal of the location credits in the LT category?

   a. Reduce vehicle travel distance

   b. Conserve a greenfield

   c. Provide diverse uses

   d. Provide surrounding density

4. Which LEED category contains prerequisites and credits?

   a. Sustainable Site

   b. Location and Transportation

   c. Innovation

   d. Regional Priority

5. Reducing the parking footprint of a building aids in what? Choose two.

   a. Provides diverse use

   b. Conserves the greenfield

   c. Reduces the heat island effect

   d. Increases walkability

6. Why is developing an urban dense area preferred? Choose three.

   a. Reduces air pollution

   b. Reduces greenhouse gas emission

   c. Conserves undisturbed areas

   d. Reduces water consumption

7. What is density?

   a. Number of buildings that could be developed in general

   b. Total building floor area or dwelling units for the buildable land of that parcel

   c. Diversity of people's background in a specific area

   d. Volume of the land

8. What is a brownfield site?

   a. A site that requires remediation to make it safe enough for occupancy

   b. A site with an active gas station

   c. A site with a building that contains asbestos

9. When used in density calculations, what does buildable land exclude? Select two.

   a. Public right of way

   b. Land excluded by codified law

   c. Infill sites

   d. Surrounding density

10. What are the two options for location strategies in the LT category?

   a. Locating the building entrance within a half-mile walking distance of seven or more diverse uses

   b. Locating the building entrance a half-mile bicycling distance of seven or more diverse uses

   c. Locating the project on or near high-density locations

   d. Locating the project within a quarter-mile radius

11. The LT credit Surrounding Density and Diverse Uses is an average density within a quarter-mile radius that depends on what?

   a. Residential density only

   b. Nonresidential density only

   c. Combination density of residential and nonresidential densities

   d. An average density of a half-mile radius of nonessential density

12. The LT credit Access to Quality Transit requires any functional entry of the project to be located

   a. within a half-mile walking distance of existing or planned bus or streetcar routes or within a quarter-mile walking distance of existing or light or heavy rail stations, commuter rail stations, or commuter ferry terminals

   b. within a quarter-mile walking distance of existing or planned bus or streetcar route or within a half-mile walking distance of existing or light or heavy rail stations, commuter rail stations, or commuter ferry terminals

   c. within a quarter-mile driving distance of existing or planned bus or streetcar routes or within a half-mile driving distance of existing or light or heavy rail stations, commuter rail stations, or commuter ferry terminals

   d. within a quarter-mile radius distance of existing or planned bus or streetcar routes or within a half-mile radius distance of existing or light or heavy rail stations, commuter rail stations, or commuter ferry terminals

13. A project team is discussing a plan for removing a solar car shading device from an existing office building. What areas may be impacted by this change?

   a. Green vehicles

   b. Sensitive land

   c. Innovation

   d. Building-level metering

14. What type of urban area is preferred by a LEED project?

   a. Area that has not been disturbed

   b. Area with zerolot lines

   c. Area with highly dense development

   d. Area on sensitive land

15. Transportation is responsible for what percentage of greenhouse gases?

   a. 20 percent

   b. 25 percent

   c. 30 percent

   d. 33 percent

16. How many points can be gained in the LT category?

   a. 10

   b. 15

   c. 16

   d. 12

17. What creates the largest environmental impact from use of a building?

   a. Transportation to and from the building

   b. Use of contaminated water

   c. Carbon footprint

   d. Sustainable Sites

18. What form of transportation is the most inefficient and most harmful to the environment?

   a. Carpooling

   b. Conventional automobiles with people traveling alone

   c. Trucks driven four times a week

   d. Vans used to drive people to and from work

19. Why does LEED encourage limited parking availability for its projects? Select two.

   a. To encourage people to carpool and use public transportation

   b. To reduce building footprints

   c. To save money and increase the return on investment

   d. To fit more building modulus in one area

20. What is the goal of the site development credits in the LT category?
    a. To provide preferred parking for carpools for 5% of the total parking spaces
    b. To promote redevelopment on existing sites over new construction on greenfield sites
    c. To protect the land and its human and wildlife occupants from future impacts
    d. To designate 5% of all parking spaces used by the project as preferred parking for green vehicles

21. What is a high-priority site?
    a. Greenfield site
    b. Brownfield site
    c. Undeveloped site
    d. Farmland

22. What is the shortest path analysis?
    a. Carpooling taking the shortest route possible
    b. Pedestrian and bicyclist travel from a point of origin to a destination
    c. Living in a residence close to work
    d. Using public transportation

23. What are the three types of boundaries?
    a. Building footprint
    b. Development footprint
    c. LEED project boundary
    d. Superfund site

24. What type of site is the historic infill district considered?
    a. Brownfield
    b. Federal empowerment zone
    c. Sensitive land
    d. High-priority site

25. What does the building footprint not include?
    a. Infrastructure
    b. Parking
    c. Roads
    d. Landscaping

26. What is the total land area of a project site covered by buildings, streets, parking areas, and other typically impermeable surfaces constructed as part of the project?
    a. Development footprint
    b. Building footprint
    c. Surrounding density and diverse use
    d. Gross area

27. Economic, environmental, and social benefits are the first key to green building that fall under which category?

    a. Materials and Resources

    b. Indoor Environmental Quality

    c. Location

    d. Energy and Atmosphere

28. Reducing greenhouse gas emission and the incidence of obesity, heart disease, and hypertension are all the intent of which category?

    a. SS (Sustainable Site)

    b. LT (Location and Transportation)

    c. MR (Materials and Resources)

    d. WR (Water and Resources)

29. What is a form of alternative transportation?

    a. Increasing the number of people traveling in a vehicle

    b. Walking

    c. A single passenger traveling in a vehicle before the rush hour

    d. Living on predeveloped lands

30. Which one is a high-priority site?

    a. Habitats

    b. Floodplains

    c. Wetlands

    d. U.S. Department of Housing and Urban Development's Qualified Census Tract (QCT) or Difficult Development Area (DDA)

31. Which are Sensitive Lands? Select two.

    a. High-priority sites

    b. Superfund sites

    c. Water bodies

    d. Farmlands

32. What is FAR?

    a. The area of ground that the building sits on as defined by its perimeter

    b. A site that has never been built on or developed for human use

    c. The density of nonresidential land use

    d. A metric for how amenable an area is to walking

33. What is NOT a diverse use in the LT category?

    a. A grocery store

    b. A library

    c. A fire station

    d. A parking lot

34. The goal of the Transportation credits is to reduce the number of people traveling to and from the building alone in conventional automobiles. This form of travel is the most inefficient and harmful to the environment. To limit the impact of transportation on the environment, what should be done? Select all that apply.

   a. Increase the use of mass transit

   b. Increase use of alternate fuel sources

   c. Increase the number of occupants per vehicle

   d. Increase parking lot capacity

35. What is the purpose of limiting parking spaces?

   a. Reducing the footprint

   b. Reducing the number of vehicles traveling to and from the site

   c. Increasing diverse use

   d. Increasing density and diverse use

36. In order to be recognized as a green vehicle, an automobile must achieve a minimum green score of _____ on the American Council for an Energy-Efficient Economy annual vehicle rating guide.

   a. 45

   b. 55

   c. 54

   d. 44

37. What is the goal of the site development credits in the LT category?

   a. Encouraging urbanism

   b. Not developing on a greenfield

   c. Protecting the land and its human and wildlife occupants from further impacts

   d. Promoting biodiversity

38. What is the measurement of how far a pedestrian and bicyclist would travel from a point of origin to a destination, reflecting access to amenities, safety, convenience, and obstructions to movement, called?

   a. Direct line of travel

   b. Shortest path analysis

   c. Offset route

   d. Detour path

   e. A simple straight-line radius

39. A city's officials are discussing a possibility of creating a diverse community. What options should they choose? Select two.

   a. Building individual residential housing

   b. Building affordable housing

   c. Building senior housing

   d. Promoting privacy by segregating affordable and senior housing

40. What is an approach that protects open space and farmland by emphasizing development with houses, jobs, and services near each other called?
    a. Smart growth
    b. Infrastructure
    c. Habitat
    d. Footprint

41. A design team is looking at a major renovation of a mall's parking lot. The parking lot has vegetated sections, and the design team wants to remove all the vegetated areas (native plants) and install a solar car shading device, which serves as a fueling station. Which of the following credits will be most impacted?
    a. Outdoor water use reduction
    b. Green vehicles use
    c. Heat island effect
    d. Site assessment

42. Building in highly dense areas reduces what?
    a. Indoor water use
    b. Outdoor water use
    c. Indoor and outdoor water use
    d. Air pollution

43. Undisturbed lands help
    a. Reduce travel distance
    b. Reduce rainwater runoff
    c. Reduce urbanism
    d. All of the above

44. How should a project team document a bicycle network for a project?
    a. Include photographs as part of the project documents
    b. Manually draw the travel path
    c. Describe in detail the bicycle network
    d. Include a vicinity map showing the shortest path analysis

45. Which one is a high-priority site?
    a. Federal empowerment zone
    b. Undisturbed site
    c. Wetland
    d. Habitat

46. How can a building facility's owner promote green vehicles?
    a. Offer a parking fee discount
    b. Designate preferred parking

    c. Building up instead of out

    d. Building the facility in an urban area

47. The project fulfilling the first credit (LEED for Neighborhood Development) in the LT category _____?

    a. Is not eligible to earn other credits in the category

    b. Gets a reduced rate on its LEED certification

    c. Is required to have one LEED AP on the project team

    d. Automatically is certified as LEED ND

48. What LEED category rewards points for dense areas and diverse uses nearby?

    a. Sustainable Site

    b. Location and Transportation

    c. Regional Priority

    d. Innovation

Reference

U.S. Census Bureau. 2021 American Community Survey: Selected Economic Characteristics. https://data.census.gov/cedsci/table?q=DP02#. Accessed date 10/4/2022.

Answers

1. **d** The Location and Transportation (LT) category is used as a guide to select a location that incurs the least environmental impact. Other intents are reducing greenhouse gas emissions, obesity, heart disease, and hypertension. Therefore, all three options (a, b, c) are important to meet the intent of the credit. Site assessment belongs to the Sustainable Site (SS) category.

2. **b**

3. **a**

4. **a**

5. **b & c**

6. **a, b, c**

7. **b**

8. **a**

9. **a & b**

10. **a & c**

11. **c**

12. **b** Option b is a requirement for Access to Quality Transit credit.

13. **a**

14. **c** The main goal of the Location and Transportation (LT) category is to select a location that reduces the amount of vehicle distance traveled to and from the building site.

15. **d**

16. **c**

17. **a** The main goal of the Location and Transportation (LT) category is to select a location that reduces the amount of vehicle distance traveled to and from the building site.

18. **b**

19. **a & b**

20. **b & c**

21. **b**

22. **b**

23. **a, b, c**

24. **d**

25. **b, c, d**

26. **a**

27. **c**

28. **b**

29. **a**

30. **d**

31. **c & d**

32. **c**

33. **d**

34. **a, b, c**

35. **b** All options are correct when looking at the Location and Transportation (LT) category in general. However, the question here is more specific about transportation and more specific about the reason for limiting parking spaces. Therefore, limiting travel to and from the site is the main aim.

36. **a**

37. **c**

38. **b**

39. **b & c** To create a diverse community, housing should be provided for a wide range of incomes and abilities. One way to promote that is to provide different housing types and not segregate them. Therefore, option a, "individual housing," and option d, "segregation," are the same thing.

40. **a** The problem statement is the definition of smart growth and also can mean redevelopment and a previously disturbed site.

41. **b & c**   Since the plants are native, there is no need for irrigation, so the first option is inapplicable. Option d belongs to the SS category. The heat island effect is impacted positively by installing a solar car shading device, as is the green vehicle for providing fueling stations.

42. **d**   Building in dense areas does not help reduce water usage.

43. **b**   If lands are left undisturbed, this may help manage rainwater runoff.

44. **d**   The shortest path analysis is a measure of pedestrian and bicyclist travel from a point of origin to a destination.

45. **a**

46. **a & b**

47. **a**

48. **b**   Diverse use near the building are the keywords to select the LT category for this particular question.

CHAPTER 5
Sustainable Sites (SS)

The Sustainable Sites (SS) category rewards decisions about the environment surrounding the building with credits that emphasize the vital relationships among buildings, ecosystems, and ecosystem services. It focuses on restoring project site elements, integrating the site with local and regional ecosystems, and preserving the biodiversity that natural systems rely on.

Earth's systems depend on biologically diverse forests, wetlands, coral reefs, and other ecosystems, which are often referred to as "natural capital" because they provide regenerative services. According to the U.N. Environment Program, the state and trends of the environment from 1987 to 2001, a U.N. study (UN, 2021) indicated that, of the ecosystem services that have been assessed worldwide, about 60% were currently degraded or used unsustainably. The results are deforestation, soil erosion, a drop in water table levels, extinction of species, and rivers that no longer run to the sea. Recent trends like exurban development and sprawl encroach on the remaining natural landscapes and farmlands, fragmenting and replacing them with dispersed hardscapes surrounded by nonnative vegetation. Between 1982 and 2001, in the United States alone, about 34 million acres (13,759 hectares) of open space (an area the size of Illinois) was lost to development—approximately 4 acres per minute or 6000 acres a day according to the U.S. Forest Service. The rainwater runoff from these hardscape areas frequently overloads the capacity of natural infiltration systems, increasing both quantity and pollution of site runoff. Rainwater runoff carries such pollutants as oil, sediment, chemicals, and lawn fertilizers directly to streams and rivers, where they contribute to eutrophication and harm aquatic ecosystems and species. A Washington State Department of Ecology study noted that rainwater runoff from roads, parking lots, and other hardscapes carries some 200,000 barrels of petroleum into the Puget Sound every year—more than half of what was spilled in the 1989 Exxon *Valdez* accident in Alaska.

Project teams that comply with the prerequisites and credits in the SS category protect sensitive ecosystems by habitat, open space, and water bodies. They use low-impact development methods that minimize construction pollution, reduce heat island effects and light pollution, and mimic natural water flow patterns to manage rainwater runoff. They also remediate areas on the project site that are already in decline.

Planting Sustainable Landscapes

Conventional plant designs and landscape maintenance practices often require irrigation and chemicals. Sustainable practices minimize the use of irrigation, fertilizers, and pesticides and can prevent soil erosion and sedimentation. Erosion from precipitation and wind causes degradation of property as well as sedimentation of local water bodies, and building sites can be major sources of sediment. Loss of nutrients, soil compaction, and decreased biodiversity of soil organisms can severely limit the vitality of landscaping. Sedimentation caused by erosion increases turbidity levels, which degrades aquatic habitats, and the buildup of sediments in stream channels can lessen flow capacity, increasing the possibility of flooding. Sustainable landscaping involves using or restoring native and adapted plants, which require less maintenance and irrigation and fewer or no applications of chemical fertilizers and pesticides compared with most introduced species. Sustainable landscaping thus reduces maintenance costs over the life of the building.

Protecting Surrounding Habitats

Development of building sites can encroach on agricultural lands and adversely affect wildlife habitat. As animals are displaced by development, they become crowded into increasingly smaller spaces; eventually, the population exceeds the carrying capacity of the area. Overall biodiversity, as well as individual plant and animal species, may be threatened. Preserving and restoring native and adapted vegetation and other ecological features on the site provide wildlife habitat.

Reducing the Heat Island Effect

The use of dark, nonreflective surface for parking areas, roofs, walkways, and other surfaces contributes to the heat island effect. These surfaces absorb incoming solar radiation and radiate that heat to the surrounding areas, increasing the ambient temperature. In addition to being detrimental to site habitat, this increase raises the building's external and internal temperature, requiring more energy for cooling. The Lawrence Berkeley National Laboratory estimated that one-sixth of the electricity consumed in the United States is used to cool buildings. By installing reflective surfaces and vegetation, the nation's homes and businesses could save $4 billion a year in reduced cooling energy demand by 2040 (U.S. Environmental Protection Agency [EPA], 2021).

Eliminating Light Pollution

Poorly designed exterior lighting may add to nighttime light pollution, which can interfere with nocturnal ecology, reduce observation of night skies, cause roadway glare, and hurt relationships with neighbors by causing light trespass. Reducing light pollution encourages nocturnal wildlife to inhabit the building site and causes less disruption to birds' migratory patterns. Thoughtful exterior lighting strategies may also reduce infrastructure costs and energy use over the life of the building.

Benefits and Issues to Consider

Environmental Issues

The loss of topsoil is the most significant on-site consequence of erosion. Topsoil is biologically active and contains organic matter and plant nutrients. Loss of topsoil greatly reduces the soil's ability to support plant life, regulate water flow, and maintain the biodiversity of soil microbes and insects that control disease and pest outbreaks. Loss of nutrients, soil compaction, and decreased biodiversity can severely limit the vitality of landscaping. This can lead to additional site management and environmental concerns such as increased use of fertilizers, irrigation, and pesticides, as well as increased storm-water runoff that adds to the pollution of nearby lakes and streams.

The off-site consequences of erosion from developed sites include a variety of water quality issues. Runoff from developed sites carries pollutants, sediments, and excess nutrients that disrupt aquatic habitats in the receiving waters. Nitrogen and phosphorus from runoff hasten eutrophication by causing unwanted plant growth in aquatic systems, including algal blooms that alter water quality and habitat conditions. Such growth can also decrease recreation potential and diminish the population diversity of indigenous fish, plants, and animals.

Sedimentation also contributes to the degradation of water bodies and aquatic habitats. The buildup of sediments in stream channels can lessen flow capacity as well as increase flooding and turbidity levels. Turbidity reduces sunlight penetration into water and leads to reduced photosynthesis in aquatic vegetation, causing lower oxygen levels that cannot support diverse communities of aquatic life.

Airborne dust from construction activity can have both environmental and human health impacts. Fine dust particles enter airways and lungs with ease and have been linked to numerous health problems, including asthma, decreased lung function, and breathing difficulties. In addition, dust particles can travel long distances before settling in water bodies, increasing the acidity of lakes and streams and changing nutrient balances.

Economic Issues

Erosion and sedimentation control measures are required by local building codes in most areas to minimize difficult and expensive mitigation measures in receiving waters. The cost will include some minimal expense associated with installing and inspecting the control measures, particularly before and after storm events, and will vary depending on the type, location, topography, and soil conditions of the project.

Implementation

Erosion typically occurs when foot traffic, runoff, or vehicle traffic damages vegetation that would otherwise hold the soil. Identifying and eliminating these and other causes will minimize soil loss and preserve receiving water quality.

The National Pollutant Discharge Elimination System (NPDES) requirements for construction activities apply only to projects of 1 acre or larger for all projects pursuing LEED (Leadership in Energy and Environmental Design) certification. Typically, the civil engineer or landscape architect identifies erosion-prone areas and outlines soil stabilization measures. The contractor then adopts a plan to implement those measures and responds to rain and other erosion-causing events accordingly. The erosion and sedimentation control plan should be incorporated into the construction drawings and specifications, with clear instructions regarding responsibilities, scheduling, and inspections. If a stormwater pollution prevention plan is required by NPDES or local regulations for the project, an erosion and sedimentation control plan may already exist. Table 5.1 shows common strategies for controlling erosion and sedimentation on construction sites.

Control Technology	Description
Stabilization	
Temporary seeding	Plant fast-growing plants to temporarily stabilize soils
Permanent seeding	Plant grass, trees, and shrubs to permanently stabilize soil
Mulching	Place hay, grass, woodchips, straw, or gravel on the soil surface to cover and hold soils
Structural Control	
Earth dike	Construct a mound of stabilized soil to divert surface runoff volumes from distributed areas into sediment basins or sediment traps
Sit fence	Construct posts with a filter fabric medium to remove sediment from stormwater volumes flowing through the fence
Sediment trap	Excavate a pond area or construct earthen embankments to allow for settling of sediment from stormwater volumes
Sediment basin	Construct a pond with a controlled water release structure to allow for settling sediment from stormwater volumes

TABLE 5.1   Strategies for Controlling Erosion and Sedimentation

Timeline and Team

During the design phase, the civil engineer or landscape architect should review local codes and create an erosion and sedimentation control plan. The general contractor should work with the project team's civil engineer or landscape architect to implement the plan during the construction phase and throughout project completion. The general contractor should photograph and maintain erosion and sedimentation control measures on-site during the various stages of construction. Once the site is stabilized, the general contractor should remove any temporary erosion and sedimentation control measures.

Summary

The LEED SS credits for New Construction, Core and Shell, and Schools promote responsible, innovative, and practical site design strategies that are sensitive to plants, wildlife, and water and air quality. These credits also mitigate some of the negative effects buildings have on the local and regional environment. Project teams undertaking building projects should be cognizant of the inherent impacts of development on land consumption, ecosystems, natural resources, and energy use. Preference should be given to buildings with high-performance attributes in locations that enhance existing neighborhoods, transportation networks, and urban infrastructures. During initial project scoping, give preference to sites and land use plans that preserve natural ecosystem functions and enhance the health of the surrounding community.

Keywords and Definitions

Floor Area Ratio (FAR): The density of nonresidential land use, exclusive of parking, measured as the total nonresidential building floor area divided by the total buildable land area available for nonresidential structures.

Green Infrastructure (GI): A soil and vegetation-based approach to wet weather management that is cost-effective, sustainable, and environmentally friendly. GI management approaches and technologies infiltrate, evapotranspire, capture, and reuse stormwater to maintain or restore natural hydrology.

Heat Island Effect: The thermal absorption by hardscape, such as dark, nonreflective pavement and buildings, and its subsequent radiation to surrounding areas.

Light Pollution: Waste light from building sites that produces glare, is directed upward to the sky, or is directed off the site. Waste light does not increase nighttime safety, utility, or security or needlessly consume energy.

Questions

The questions and multiple-choice answers listed next were designed to provide a deeper insight and look into the SS category by addressing the following:

1. What credit rewards projects for site ecosystems?
   a. Sustainable Sites (SS)
   b. Water Efficiency (WE)
   c. Innovation (IN)
   d. Location and Transportation (LT)

2. What does the SS Prerequisite: Construction Activity Pollution Prevention aim to reduce? Choose three.
   a. Airborne dust
   b. Waterway sedimentation
   c. Soil erosion
   d. Increased density
   e. Heat reduction

3. The SS Credit: Site Development—Protect and Restore Habitat encourages project teams to designate areas as protected habitat and open space for the life of the project. What helps to maintain overall ecosystem health? Select two.
   a. Installing invasive plants
   b. Reducing greenhouse gas emission
   c. Restoring native plants
   d. Restoring native soils

4. What are the two healthcare credits that build off the idea of providing open space?
   a. Sensitive land protection
   b. Quality view

    c. Minimum indoor air quality performance

    d. Places of respite

    e. Direct exterior access

5. For a building in an arid climate, its team is working to reduce the heat island effect. What should the team do?

    a. Enhance the roof with more insulation

    b. Eliminate unnecessary daylighting

    c. Use high-reflectance materials for the entire roof

    d. Use a 15 percent green roof

6. List two of the below that promote biodiversity:

    a. Athletic field

    b. Playground with artificial turf

    c. Undisturbed area

    d. Intensive and extensive green roof

7. What are the luminaire classifications of light pollution?

    a. Backlight and uplight

    b. Glare

    c. Forward light high

    d. Motion sensors and timers

8. Locating parking underground and covering the roof using native plants achieve which credit categories? Choose two.

    a. Sensitive Land

    b. Heat Island Effect

    c. Open Space

    d. Access to Quality Transit

9. What does a LEED project define open space as?

    a. Grassy playground that has no access to users

    b. A building roof that is covered by native and adaptive plants, but hard to maintain

    c. 50 percent of the project site that is open to public for physical activities and social interaction

    d. Space within the LEED project boundary that is open for credit interpretation

10. The solar reflectance index is a metric used to measure how well a material reflects solar heat. How should the project team decide on selecting a material for the project for better performance?

    a. Low-reflective materials

    b. High-reflective materials

    c. High absorption of heat

    d. Low absorption of heat

11. A land management strategy that emulates a natural system is to manage rainwater as close to its source as possible. What is this the definition of?

    a. Floor area ratio (FAR)

    b. Albedo

    c. Low-impact development (LID)

    d. Watershed

12. SS Credit: Site Development—Protect or Restore Habitat encourages project teams to designate areas as protected habitat and open space for the life of the project. On-site restoration helps to maintain overall ecosystem health by restoring what? Choose three.

    a. Native soils

    b. Plants

    c. Hydrology

    d. Invasive plants

13. The main aim of Construction Activity Pollution Prevention is to control what? Choose three.

    a. Soil erosion

    b. Waterway sedimentation

    c. Generation of airborne dust

    d. Reduced building footprint size

14. LEED projects encourage increasing site density by having which of the following?

    a. Higher floor area ratio (FAR).

    b. Lower floor area ratio.

    c. Midvalue floor area ratio.

    d. Floor area ratio is not applicable to site density.

15. What does the SS Credit: Light Pollution Reduction use to select luminaires?

    a. High BUG (backlight, uplight, and glare) ratings

    b. Low BUG ratings

    c. BUG ratings are not relevant to the Light Pollution Reduction credit

16. What is the area on a project site that is used by the building structure, defined by the perimeter of the building plan, called?

    a. Place of respite

    b. Building footprint

    c. Project perimeter

    d. Building surrounding

17. What is the main consideration in the SS category of the BD+C (Building Design and Construction) rating system? Select two.

    a. Cooling tower blowout

    b. Selection of a building site and managing that site during construction

     c. Consideration for environmental impact over its lifetime

     d. Xeriscaping

18. What are asphalt, concrete, brick, stone, and sealed surface examples of?

     a. Pervious area

     b. Impervious area

     c. Floor area ratio (FAR)

     d. American Society of Heating, Refrigerating, and Air-Conditioning (ASHRAE)

19. Exterior space that encourages interaction with the environment, social interactions, passive recreation, and physical activities is called what?

     a. Close space

     b. Open space

     c. Indoor space

     d. On-site solar panel renewal

20. What are two benefits when using pervious materials for hardscapes?

     a. Reduction of rainwater runoff

     b. Albedo

     c. Restoration of the local water table

     d. Nonpoint source pollution

21. What are the two environmental impacts you achieve when installing underground parking?

     a. Rainwater runoff

     b. Increased convenience

     c. Heat island effect

     d. Shortening of scheduled activities

22. What is a cistern an example of?

     a. Active rainwater management

     b. Rainwater runoff

     c. Water table management

     d. Building footprint

23. What is redirecting rainwater to planted areas where it is allowed to saturate the soil is of?

     a. Reduced impervious hardscape areas

     b. Passive rainwater management

     c. Environmental site assessment

     d. Heat island reduction

24. What definition applies to a metric from 0 to 100 that measures how well a material reflects solar heat, with higher numbers signifying reflectance?

    a. Ecosystem

    b. Heat island effect

    c. Solar Reflectance Index (SRI)

    d. Albedo

25. Which credit falls under the SS category?

    a. Surrounding density and diverse use

    b. Open space

    c. Water metering

    d. Demand response

26. Reducing the environmental impact of developing a building site and maintaining it for the life of the building is the goal of which category?

    a. LT

    b. SS

    c. Energy and Atmosphere (EA)

    d. Innovation

27. What is the intent of rainwater management?

    a. To guide the construction management of the project throughout its lifetime to limit environmental destruction inherent to building

    b. To reduce rainwater runoff and improve water quality by duplicating the natural hydrology and water balance of the site

    c. To reduce heat islands and minimize their effects on regional climates and human health and wildlife habitats

    d. To protect the land and its human and wildlife occupants from further impacts

28. What is the primary goal of SS Prerequisite: Construction Activity Pollution Prevention?

    a. Promoting redevelopment on existing sites over new construction on greenfield sites

    b. Reducing the impact of pollution on the building site due to construction activity

    c. Providing rainwater management

    d. None of the above

29. Name one building method that can help increase the floor area ratio.

    a. Using green infrastructure (GI)

    b. Building up rather out

    c. Providing total parking capacity

    d. Building out rather up

30. What is the definition of an increase in microclimate temperature created by waste heat from human activity, building operation, and surfaces in the built environment that absorbs sunlight?

    a. Carbon offset

    b. British thermal unit

    c. Energy use intensity (EUI)

    d. Heat island effect

31. On-site assessment (Phase 1) is to determine whether environmental contamination exists. If Phase 2 assessment indicates soil or groundwater contamination, site remediation must occur. What do these strategies fall under?

    a. Construction activity pollution prevention

    b. Site assessment

    c. Environmental site assessment

    d. Building Product Disclosure and Optimization—Environmental Product Declarations

32. Select three methods to reduce the heat island effect.

    a. Installing a green roof

    b. Using open-grid paving

    c. Providing shade for hardscapes

    d. Installing Hardscape

    e. Installing Impervious pavement

33. What are some rainwater management strategies? Select three.

    a. Providing green infrastructure (GI)

    b. Using low-impact development (LID)

    c. Preserving ecological integrity and encouragement of environmentally sensitive site management practices

    d. Providing outdoor space greater than or equal to 30 percent of the total site area

    e. Reducing the amount of impervious areas

34. Construction pollution, protection and restoration of habitat, and reduction of the size of the building footprints are all strategies of what?

    a. Site design and management

    b. Maximization of open spaces

    c. Passive rainwater management

    d. Soil erosion prevention

35. What is the definition of the land management strategy that emulates natural systems to manage rainwater as close to its source as possible?

    a. Green infrastructure (GI)

    b. Low-impact development (LID)

c. Integrated pest management (IPM)

d. Development density (DD)

36. What does the practice of reducing or eliminating potable water use in irrigation through the planting of native and adapted species or vegetation refer to?

a. Water Policy Act of 1992

b. WaterSense

c. Xeriscaping

d. Aeration

37. What does a well-balanced diverse ecosystem provide?

a. Controlled erosion

b. Albedo

c. Clear air and water

d. Bioswale

38. Sediment carried by rainwater runoff from construction sites, documented as one of the leading sources of pollution to streams and rivers, can be prevented with implementing what?

a. Enlarging the building footprint

b. Having a construction activity pollution plan

c. Implementing Phase I (Environmental Site Assessment)

d. Implementing Phase I and Phase II (Environmental Site Assessment)

39. Building up instead of out contributes to which of the following?

a. A gain of 6 points on the Innovation category

b. Surrounding density and diverse use

c. Building life cycle impact reduction

d. Open space

40. Reducing the construction impact to the land, reducing building materials systems, reducing the costs of mechanical systems and the cost of heating air conditioning, and ventilating the building all can be achieved by which process?

a. Building orientation

b. Site development

c. Construction activity pollution prevention

d. A smaller footprint

41. Select all applicable examples of pervious surfaces for sidewalk and pavement.

a. Vegetated roof

b. Porous pavement

c. Asphalt

d. Grid pavers

42. A passive rainwater-harvesting system is an inexpensive, yet effective, method of controlling rainwater runoff and reducing soil erosion. Select all applicable examples of passive rainwater systems.

    a. Drainage pipes to direct water to areas within the building

    b. Swales

    c. Berms

    d. Permeable pavement

43. What is the method of a collection of rocks or boulders held together using wire mesh or metal fabric materials for the purpose of reducing soil erosion called?

    a. Erosion prevention

    b. Flood stoppage

    c. A gabion

    d. Rainwater runoff approach

44. What is created by waste heat from human activity, building operation, and surfaces in the built environment that absorb sunlight?

    a. Heat island effect

    b. British thermal unit (Btu)

    c. Carbon waste

    d. ASHRAE 90.1-2021

45. Select two approaches that contribute to reducing the heat island effect:

    a. Reduction of exposed hardscape

    b. Use of green power and carbon offsets

    c. Use of high-reflectance materials

    d. Attention to development density (DD)

46. Rainwater management has synergies with which of the following?

    a. Construction pollution prevention

    b. Floor area ratio (FAR)

    c. A dry pond

    d. Light pollution

47. In a desert area, an owner wants to reduce light pollution for his or her facility. What is the best practice?

    a. Installing light that shines sideward

    b. Using open-grid paving

    c. Installing lights that reduce glare

    d. Installing lights with a high solar reflectance index

48. To promote biodiversity, which option should a team choose? Select all that apply.

    a. Financial support of a conservation organization

    b. Green roof

c. Football field

d. A playground with artificial grass

49. A project team is looking at options for eliminating light pollution and conserving energy for a football field's parking garage. What should they do?

a. Install lights by staggering them to ensure light coverage for the entire parking garage

b. Install lights that can shine backward and upward

c. Install lights with timers

d. None of the above

50. What are the best ways to mimic nature and manage rainwater?

a. Using xeriscaping

b. Having green infrastructure (GI)

c. Using low-impact development (LID)

d. Using grid pavers

51. Open space logically contributes to what? Select all that apply.

a. Managing rainwater

b. Reducing the heat island effect

c. Reducing light pollution

d. Providing integrated pest management (IPM)

52. A building owner needed advice from a LEED professional on the type of plants that should be used on a proposed vegetated roof. Which option is the best choice?

a. Trees

b. Tall grass and evergreens

c. Indigenous plants

d. Shrubs

53. A team is designing an open space for a project. Which option is the most appropriate?

a. A place for sport activities

b. A private indoor garden

c. A football field next door

d. All of the above

54. To reduce the heat island effect, what should the selection of materials to pave a sidewalk and cover a roof be based on?

a. The thickness of the materials

b. The darkness of the materials

c. Materials that should have high solar reflectance

d. Materials with a high solar reflectance index

References

U.S. Environmental Protection Agency (EPA). (2021). Reduce **Urban Heat Island Effect**. https://www.epa.gov/green-infrastructure/reduce-urban-heat-island-effect. Accessed January 2022.

UN (2021), environment programme. Emissions Gap **Report: The Heat Is On**. https://wedocs.unep.org/bitstream/handle/20.500.11822/37287/EGR21AP.pdf. Accessed 2022.

Answers

1. **a**   The SS category encourages responsible site design and management because a building and its site can have a large impact on the environment.

2. **a, b, c**

3. **c & d**   Option c uses less irrigation and option d maintains the ecosystem environment (i.e., an ecosystem is a biological community of interacting organisms and their physical environment).

4. **d & e**   Option a is for the LT category, options b and c are for IAQ. Options d and e are for healthcare credits.

5. **c**   Option b is irrelevant. For the rest of the options, this question is tricky, so all options are applicable. However, which option is the best option? Option c is the best option for reflecting the heat. Option a, adding more insulation, uses more money and is not necessarily an effective approach. Option d is not very effective.

6. **c & d**   Options c and d encourage biodiversity, while the other options do not.

7. **a & b**   Check the classifications for light pollution.

8. **b & c**   Options b and c address the SS category; the other options address the LT category.

9. **c**

10. **b**   This option helps reduce heat island emissions.

11. **c**

12. **a, b, c**

13. **a, b, c**

14. **a**   FAR is the concept of building up rather than out and maximizes the building footprint without decreasing the square feet.

15. **b**   Backlight, uplight, and glare (BUG) ratings are used to classify luminaires and their likelihood of generating light pollution.

16. **b**

17. **b & c**   Options a and d are more for the Water efficiency (WE) category, so b and c options are suitable.

18. **b**

19. **b**   The problem statement defines the intent of the open space.

20. **a & c**

21. **a & c**

22. **a**

23. **b**

24. **c**

25. **b**

26. **b**

27. **b**

28. **b**

29. **b**

30. **d**

31. **c**

32. **a, b, c**

33. **a & b**   Option d is irrelevant, and it is more for open space requirements. Option c is more for the LT category and sensitive lands.

34. **a**

35. **c**

36. **c**

37. **a & c**   Option b is more for surface reflection, and option d is a feature, not a provided benefit.

38. **b**   The construction activity pollution plan aims to reduce pollution from construction activities by controlling soil erosion, waterway sedimentation, and the generation of airborne dust.

39. **d**   Reducing the footprint by building up instead of out contributes to maximizing the open space of the property.

40. **d**   Building orientation helps mostly in natural ventilation and lighting/heating. Site development helps mainly with the impact on the land. Construction activity pollution prevention helps with soil erosion and waterway sedimentation. Therefore, option d, a smaller footprint, is the best solution.

41. **b, c, d**

42. **b & c**   Options b and c are examples of passive rainwater-harvesting systems, which are hardscape features designed to clean and hold rainwater and drain it back into the earth.

43. **c**   Option c is the correct answer. A gabion slows the movement of water, which allows for saturation into the soil.

44. **a**

45. **a & c**

46. **a**   Construction pollution prevention aims to control soil erosion.

47. **c**   Lights should shine downward, not upward and sideward. The other options (b and d) are irrelevant.

48. **a & b** Options c and d do not promote biodiversity because both use artificial grass. Option b promotes biodiversity, and option a promotes habitat conservation and restoration, which as a result promotes biodiversity.

49. **c** Option c is the best option. Option b is incorrect because lights are not supposed to shine upward. Option a is incorrect because staggering lights causes energy consumption.

50. **b & c** Green infrastructure is defined as the patchwork of natural areas that provide habitat, flood protection, clean air, and clean water at the scale of a city or county or rainwater management systems that mimic nature by soaking up and storing water at the scale of a neighborhood or site. Low-impact development is defined as a land management strategy that emulates natural systems to manage rainwater as a source of irrigation.

51. **a & b** Options c and d are irrelevant. When open space is designed properly, it can be used to manage rainwater and reduce the heat island effect.

52. **c** Indigenous plants are plants that grow in the same area and are native plants that require less water.

53. **a** The intent of open space is to encourage interaction with the environment, social interaction, passive recreation, and physical activities. The football field is good, but it should be within the project's boundaries, not next door. Open space should be accessible, so a private garden is not applicable.

54. **c & d** The aim is to reflect sunlight and reduce absorption of heat from the sun to reduce the heat island effect.

CHAPTER 6

Water Efficiency (WE)

The "Water Efficiency" (WE) chapter addresses water holistically, looking at indoor use, outdoor use, specialized uses, and metering. The chapter is based on an "efficiency-first" approach to water conservation. As a result, each prerequisite looks at water efficiency and reduction in potable water use alone. Then, the WE credits additionally recognize the use of nonpotable and alternative sources of water.

The U.S. Environmental Protection Agency (EPA, 2018) stated that the conservation and creative reuse of water are important because only 3 percent of Earth's water is fresh water, and of that slightly over two-thirds are trapped in glaciers. Typically, most of a building's water cycles through the building and then flows off-site as wastewater. In developed nations, potable water often comes from a public water supply system and then flows off-site; wastewater leaving the site must be piped to a processing plant, after which it is discharged into a distant water body. This pass-through system reduces streamflow in rivers and depletes freshwater aquifers, causing water tables to drop and wells to go dry. Based on The United Nations (U.N.) statistics as shown on graphs and maps, in 60 percent of European cities with more than 100,000 people, groundwater is being used faster than it can be replenished (UN water).

In addition, the energy required to treat water for drinking, transport it to and from a building, and treat it for disposal represents a significant amount of energy use not captured by a building's utility meter. Research in California showed that roughly 19 percent of all energy used in this U.S. state was consumed by water treatment and pumping, with the information retrieved from California's energy website. Green building facts from the U.S. Green Building Council indicated that in the United States buildings accounted for 13.6 percent of potable water use, the third-largest category, behind thermoelectric power and irrigation. Designers and builders can construct green buildings that use significantly less water than conventional construction by incorporating native landscapes that eliminate the need for irrigation, installing water-efficient fixtures, and reusing wastewater for nonpotable water needs. The Green Building Market Impact Report (TGBMIR, 2009) found that LEED (Leadership in Energy and Environmental Design) projects were responsible for saving an aggregate 1.2 trillion gallons (4.54 trillion liters) of water. LEED's WE credits encourage project teams to take advantage of every opportunity to significantly reduce total water use.

The water category comprises three major components: indoor water (used by fixtures, appliances, and processes, e.g., cooling); irrigation water; and water metering. Several kinds of documentation span these components, depending on the project's specific water-saving strategies.

Documentation Needed

Site Plan

Plans are used to document the location and size of vegetated areas and the locations of meters and submeters. Within the building, floor plans show the location of fixtures, appliances, and process water equipment (e.g., cooling towers, evaporative condensers), as well as indoor submeters. The same documentation can be used in credits in the Sustainable Site (SS) category.

Fixture Cut Sheets

Projects must document their fixtures (and appliances as applicable) using fixture cut sheets or manufacturer's literature. This documentation is used in the Indoor Water Use Reduction prerequisite and credit.

Alternative Water Source

A project that includes graywater reuse, rainwater harvesting, municipally supplied wastewater (purple pipe water), or other reused sources is eligible to earn credit in WE Credit Outdoor Water Use Reduction, WE Credit Indoor Water Use Reduction, WE Credit Cooling Tower Water Use, and WE Credit Water Metering. But the team cannot apply the same water to multiple credits unless the water source has sufficient volume to cover the demand of all the uses (e.g., irrigation plus toilet-flushing demand).

Occupancy Calculation

The Indoor Water Use Reduction prerequisite and credit require projections based on occupants' usage. The Location and Transportation and SS categories also use project occupancy calculations. Review the occupancy section in getting started to understand how occupants are classified and counted. Also see Appendix A for more on water efficiency and water calculations. The appendix includes calculations for real-world projects.

Water Consumption

Americans' consumption of the public water supply continues to increase. The U.S. Geological Survey estimated that between 1990 and 2000, this consumption increased 12 percent to 43.3 billion gallons per day. The public water supply is delivered to users for domestic, commercial, industrial, and other purposes and is the primary source of water for most buildings. The U.S. Geological Survey report of 2008 showed that these uses represented about 11 percent of total withdrawals and slightly less than 40 percent of groundwater withdrawals, constituting the third-largest category of water use in the United States, behind thermoelectric power (48 percent of total withdrawals) and irrigation (34 percent of total withdrawals). This high demand for water is straining suppliers, and in some parts of the United States, water levels of underground aquifers have dropped more than 150 feet since the 1940.

Only about 14 percent of withdrawn water is lost to evaporation or transportation or incorporated into production of crops; the rest is used, treated, and discharged to the nation's water bodies (Solley et al., 1998). Discharged water contaminates rivers, lakes, and potable water with bacteria, nitrogen, toxic metals, and other contaminants. The EPA estimated that one-third of the nation's lakes, streams, and rivers are not safe for swimming and fishing. Even so, water bodies in the United States are 50 percent cleaner today than in the mid-1970s. And although consumption is rising, total U.S. withdrawals from the public water supply declined by nearly 9 percent between 1980 and 1985 and have varied by less than 3 percent for each 5-year interval since then (Maupin et al., 2010).

Using large volumes of water increases maintenance and life-cycle costs for building operations and also increases consumers' costs for the additional municipal supply and treatment facilities (Alshareef et al., 2020). Conversely, buildings that use water efficiently can reduce costs through lower fees, reduced volume of sewage, reductions in energy and chemical use, and lower capacity changes and limits.

Efficiency measure can easily reduce water use in average commercial buildings by 30 percent or more (U.S. Green Building Council, 2021). In a typical 100,000-square-foot office building, low-flow plumbing fixtures coupled with sensors and automatic controls will save a minimum of 1 million gallons of water per year (Allen et al., 2016). In addition, no potable water can be used for landscaping irrigation, toilet and urinal flushing, custodial purposes, and building systems. Depending on local water costs, utility savings can be tens of thousands of dollars per year. The real estate firm Cushman and Wakefield, for example, implemented a comprehensive water management strategy at its adobe headquarters in San Jose, California, in 2002 and achieved a 22 percent reduction in water use (Knox, 2007).

The LEED for New Construction, LEED for Core and Shell, and LEED for Schools Water Efficiency prerequisite and credits encourage the use of strategies and technologies that reduce the amount of potable water consumed in buildings. Many water conservation strategies are no cost or provide a rapid payback. Other strategies, such as biological wastewater treatment systems and graywater plumbing systems, often require more substantial investment and are cost-effective only under certain building and site conditions.

The WE prerequisite and credits address environmental concerns relating to building water use and disposal and promote the measures that follow.

Monitoring Water Consumption Performance

The first step in improving WE is to understand current performance. Tracking water use alongside energy use can help organizations better understand how these resources relate to each other, make integrated management decisions that increase overall efficiency, and verify savings from improvement projects in both energy and water systems. Organizations that manage water and energy performance together can take advantage of this relationship to create greener, more sustainable buildings.

Reducing Indoor Potable Water Consumption

Reducing indoor potable water consumption may require using alternative water sources for nonpotable applications and installing building composting toilet systems and waterless urinals. Lowering potable water use for toilets, showerheads, faucets, and other fixtures can reduce the total amount withdrawn from natural water bodies. A commercial building in Boston replaced 126 toilets using 3.5 gallons per flush (gpf) with low-flow, 1.6-gpf toilets and reduced total water use by 15 percent. With an initial cost of $32,000 and an estimated annual savings of $22,800, the payback from the renovation was 1.4 years. Another Boston building installed 30 faucet aerators and reduced annual indoor water consumption by 190,000 gallons. The cost of the materials and labor totaled $300, and the change is estimated to save $1250 per year, with a simple payback of 2 months (Massachusetts Water Resources Authority, 2021).

Reducing Water Consumption to Save Energy and Improve Environmental Well-Being

In many buildings, the most significant savings associated with WE result from reduced energy costs. WE cuts costs by reducing the amount of water that must be treated, heated, cooled, and distributed, all of which require energy. According to the commercial energy consumption survey (CBECS, 2018), significant energy savings comes through efficient use of hot water because water heating in commercial buildings accounts for nearly 15 percent of total building energy use. For this reason, water conservation that reduces the use of hot water also conserves energy and reduces energy-related pollution. For example, U.S. government office buildings use an estimated 244 to 256 billion gallons of water each year. Approximately 138.3 billion Btu of energy are required to process this water annually, 98 percent of which is used to heat water. By implementing water-efficiency efforts, federal buildings could conserve approximately 40 percent of total water consumption and reduce related energy use by approximately 81.32 billion Btu per year.

Practicing water conservation measures can also help improve both environmental and human well-being. A recent U.S. government survey showed that at least 36 states are anticipating local, regional, or statewide water shortages by 2030 (EPA, 2018). Human health and environmental welfare are affected when reservoirs and groundwater aquifers are depleted because lower water levels can concentrate both natural contaminants such as radon and arsenic and human pollutants such as agricultural and chemical wastes. Increasing WE helps keep contaminants at safe levels.

Water efficiency also reduces energy consumption in the water supply and wastewater infrastructure. American public water supply and treatment facilities consume about 56 billion kilowatthours (kWh) each year (EPA, 2018), enough electricity to power more than 5 million homes for an entire year. Better WE in commercial buildings will reduce the amount of energy consumed by water treatment facilities.

Practicing Water-Efficient Landscaping

Landscaping irrigation practices in the United States consume large quantities of potable water. Outdoor uses, primarily landscaping, account for 30 percent of the 26 billion gallons of water consumed daily (EPA, 2018). Improved landscaping practices can dramatically reduce and even eliminate irrigation needs. Maintaining or reestablishing native plants on building sites fosters a self-sustaining landscape that requires minimal supplemental water and provides other environmental benefits.

Native plants require less water for irrigation and attract native wildlife, thus creating a building site integrated with its natural surroundings. In addition, native plants tend to require less fertilizer and pesticides, avoiding water quality degradation and other negative environmental impacts.

In Schools, Use Water-Efficient Processes as a Teaching Tool

Many systems used for WE provide a wealth of educational opportunities, including the study of biological systems, nutrient cycles, habitats, and the impact of human systems on local watersheds and natural resources. Students can calculate the effects of water conservation strategies on their own water use, simultaneously practicing math skills and environmental stewardship. Schools that have constructed wetlands or rain collection and distribution systems can consider making these technologies highly visible components of the school design.

Employ strategies that in aggregate use 20 percent less water than the water use baseline calculated for the building (not including irrigation).

Calculate the water consumption baselines for commercial usages according to Table 6.1. Also, calculate the water consumption baselines for residential usage according to Table 6.2. Calculations are based on estimated occupant usage and must include only the following fixtures and fixture fittings (as applicable to the project scope): water closets, urinals, lavatory faucets, showers, kitchen sink faucets, and prerinse spray valves.

Commercial Fixtures, Fittings, and Appliances	Current Baseline
Commercial toilets	1.6 gpf, except blowout fixtures (3.5 gpf)
Commercial urinals	1.0 (gpf)
Commercial lavatory (restroom) faucets	2.2 gallons per minute (gpm) at 60 pounds per square inch (psi), private applications only (hotel or motel guest rooms, hospital patient rooms)
	0.5 gpm at 60 psi at other applications except private applications
	0.25 gallons per cycle for metering faucets
Commercial prerinse spray valves (for food service applications)	Flow rate ≤ 1.6 gpm
	(no pressure specified; no performance requirement)

TABLE 6.1   Commercial Water Consumption Baselines for Fixtures, Fittings, and Appliances

Commercial Fixtures, Fittings, and Appliances	Current Baseline
Residential toilets	1.6 gpf
Residential lavatory (bathroom) faucets	2.2 gpm at 60 psi
Residential kitchen faucet	
Residential showerheads	2.5 gpm at 80 psi per shower stall

Note: Tables 6.1 and 6.2 were adapted from information developed and summarized by the U.S. Environmental Protection Agency (EPA) office of water based on requirements of the Energy Policy Act (EPAct) of 1992 and subsequent ruling by the Department of Energy, requirements of the EPAct of 2005, and the plumbing code requirements as stated in the 2010 editions of the uniform plumbing code or International Plumbing Code pertaining of fixture performance.

TABLE 6.2   Residential Water Consumption Baselines for Fixtures, Fittings, and Appliances

Benefits and Issues to Consider

Environmental Issues

Reducing potable water use in buildings for urinals, toilets, showerheads, and faucets decreases the total amount withdrawn from rivers, streams, underground aquifers, and other water bodies. The strategies protect the natural water cycle and save water resources for future generations. In addition, water use reductions, in aggregate, allow municipalities to reduce or defer the capital investment needed for the water supply and wastewater treatment infrastructure.

Conserving municipally supplied potable water also reduces chemical inputs at the water treatment works, as well as reduces energy use and the associated greenhouse gas emissions from treatment and distribution. The energy use and emissions generated to supply municipal water vary greatly across the United States and depend on the utility's water sources, the distances water is transported, and the type of water treatment applied. End-use WE can greatly reduce negative environmental impacts. Comparing the environmental effects of off-site treatment and supply with those of on-site treatment is a worthwhile exercise. Because water heating in commercial buildings accounts for nearly 15 percent of building energy use, conservation measures will also reduce end-use energy and energy-related pollution.

Economic Issues

Reduction in water consumption decreases building operating costs and brings about wider economic benefits. Reduced water consumption allows municipalities to lessen or defer the capital investment needed for water supply and wastewater treatment infrastructure, thereby leading to more stable municipal taxes and water rates.

Many cost-effective systems and fixtures currently on the market support compliance with the requirements, but the cost of WE measures varies widely. For example, installing tamperproof faucet aerators on exiting fixtures is a small expense compared with a rainwater-harvesting or graywater-recycling system. High-efficiency toilets and dry fixtures such as nonwater toilet systems often have higher initial costs than standard models.

Newer technologies may also have higher costs and limited availability because of production constraints, and they may entail different maintenance and repair expenses, such as special cartridge components and cleaning and sealing fluids. Teams should perform a full cost-benefit and life-cycle study before installing such products.

Referenced Standards

EPAct of 1992 (and as Amended)

The U.S. EPAct of 1992 addresses energy and water use in commercial, institutional, and residential facilities.

EPAct of 2005

The EPAct statute became U.S. law in August 2005.

International Association of Plumbing and Mechanical Officials Publication IAPMO/American National Standards Institute UPC1-2006, Uniform Plumbing Code 2006, Section 402.0, Water-Conserving Fixture and Fittings

The Uniform Plumbing Code of 2006 defined water-conserving fixtures and fittings for waste water closets, urinals, and metered faucets. This code accredited by the American National Standards Institute safeguard life, health, property, and public welfare by regulating and controlling the design, construction, installation, materials, location, operation, and maintenance or use of plumbing systems.

International Code Council, International Plumbing Code 2010, Section 604, Design of Building Water Distribution System

The International Plumbing Code of 2010, Section 604 defines maximum flow rates and consumption for plumbing fixtures and fittings, including public and private lavatories, showerheads, sink faucets, urinals, and water closets.

Implementation

Effective ways to reduce water use include installing flow restrictors or reduced-flow aerators on lavatory, sink, and shower fixtures; installing and maintaining automatic faucet sensors and metering controls; installing low-consumption flush fixtures such as high-efficiency water closets and urinals; installing nonwater fixtures; and collecting rainwater.

In certain cases, faucets with low flow rates are not appropriate. For example, in kitchen sinks and janitors' closets, faucets are used to fill pots and buckets. Using a low flow rate for tasks where the volume of water is predetermined does not save water and will likely cause frustration. Consider alternative strategies to reduce water use, such as installing special-use pot fillers and high-efficiency faucets or faucets operated by a foot pedal.

WaterSense, a partnership program sponsored by the EPA helps consumers identify water-efficient products and programs. WaterSense-labeled products exceed the Uniform Plumbing Code and the International Plumbing Code standards for some high-efficiency fixtures or fittings; WaterSense products and other high-efficiency plumbing fixtures, fittings, and appliances can be installed in the same way as conventional EPAct plumbing fixtures, fittings, and appliances.

To determine the most effective strategies for a particular condition, analyze the water conservation options available to the project based on location, code compliance (plumbing and safety), and overall project function. Determine where in the building the most water is used, evaluate potential alternative water-saving technologies, and examine the impacts of alternative fixtures and technologies. Compare the water use of the proposed design with the calculated EPAct baseline to determine the optimal water savings for plumbing fixtures and fittings. Once the design's water use has been determined, compare the volumes of water required for each end use with the volumes of alternative sources of water available on site. Perform a detailed climate analysis to determine the availability of on-site resources and choose strategies that are appropriate and cost-effective.

Timeline and Team

During predesign, setting water goals and strategy involves the owner, architect, and engineers. Identify local water utilities and governing authorities and research codes and applicable water laws. Learn the process for obtaining permits and approval and set water goals and strategy.

During design development, the engineering team should develop and design water reuse and treatment systems, perform preliminary LEED calculations, and confirm or reassess water goals.

In construction documents, the architect, working with the owner, should specify efficient fixtures and appliances and complete LEED calculations and documentation.

Keywords and Definitions

Aquifer: An underground water-bearing rock formation or group of formations that supply groundwater, wells, or springs.

Biochemical Oxygen Demand: A measure of how fast biological organisms use oxygen in a body of water. The measure is used in water quality management and assessment, ecology, and environmental science.

Black Water: In a sanitation context, black water denotes wastewater from toilets, which likely contains pathogens that may spread by the fecal-oral route. Black water can contain feces, urine, water, and toilet paper from flush toilets.

Graywater: Domestic wastewater generated in households or office buildings from streams without fecal contamination (i.e., all streams except for the wastewater from toilets). Sources of graywater include sinks, showers, bathtubs, washing machines, or dishwashers.

Potable Water: Potable water meets or exceeds the EPA's drinking water quality standards and is approved for human consumption by the state or local authorities having jurisdiction; it may be supplied from wells or municipal water systems.

Process Water: Water used for industrial processes and building systems such as cooling towers, boilers, and chillers. The term can also refer to water used in operational processes, such as dishwashing, clothes washing, and ice making.

Questions

During construction, the design team and owner should confirm proper selection, installation, and operation of water fixtures, fittings, and systems. The questions and multiple-choice answers that follow were designed to provide a deeper insight and look into the WE category.

1. What is the baseline water consumption for a public lavatory faucet?

  a. 0.3 gpf

  b. 0.5 gpm

  c. 0.5 gpf

  d. 1.6 gpm

2. To reduce outdoor water consumption, what type of plants should be used? Choose two.

   a. Oak tree

   b. Turf grass

   c. Native plants

   d. Adaptive plants

3. A team is deciding on the potable water selection for a project. Which one is potable water?

   a. Open-loop geothermal water

   b. Cooling tower blowout

   c. Municipally recycled water used for landscape irrigation

   d. Well water that meets the EPA's drinking standards

4. What does reducing indoor water use strategies include?

   a. Implementing a graywater or rainwater reclamation system

   b. Harvesting rainwater

   c. Using sprinklers and drip irrigation

   d. Using xeriscaping

5. A project team wants to justify the additional cost of adding submetering for a retail project. What should the team tell the owner?

   a. Submetering is one of the Minimum Program Requirement.

   b. Submetering helps users measure their consumption in regard to how the system operates and performs.

   c. Submetering is required during the discovery phase.

6. What is the intent of the WE category?

   a. Increase nonpotable water use

   b. Limit the use of drinking water to bathrooms and kitchens only

   c. Decrease potable water use

   d. Decrease water used in buildings for nonpotable water uses

7. Which standards does the WE category use to measure water consumption from the baseline?

   a. American Society of Heating, Refrigeration and Air-Conditioning Engineers (ASHRAE) 62.1

   b. ASHRAE 55.9

   c. EPAct of 1992

   d. EPA portfolio manager

8. A community center has experienced an increase in water usage due to users' high demands. Which one of the following strategies would help in water reduction?

    a. Install dual-flush water closets

    b. Replace the turf grass with xeriscaping

    c. Install battery-based dispensers

    d. Change the building orientation

9. What are the water measurement units by usage for faucets and showers?

    a. Gallons per flush (gpf).

    b. Gallons per minute (gpm).

    c. Gallons per hour (gph).

    d. Once the baseline is measured, there is no need for measuring the design case value.

10. How many prerequisites are in the WE category?

    a. Two

    b. Three

    c. Four

    d. Five

11. Which technology may NOT help the project team reduce indoor water use?

    a. Water reclamation system

    b. Composting toilet

    c. Dual-flush water closet

    d. More efficient irrigation system

12. What is water that comes from showers, baths, lavatory sinks, and clothes washers called?

    a. Black water

    b. Gray water

    c. Process water

    d. Reclaimed water

13. What is measuring or projecting the indoor water use based on?

    a. Occupants' usage

    b. Number of water fixtures in the building

    c. Total square footage of the building

    d. Cost of water per gallon

14. What is the baseline water consumption for a water closet (toilet)?

    a. 0.3 gpf

    b. 0.5 gpm

c. 0.5 gpf

d. 1.6 gpm

15. A project team is deciding on utilizing gray water in a commercial building. This practice would help reducing potable water in which of the options below?

   a. Commercial dishwashers.

   b. Urinals.

   c. Laundry equipment.

   d. Gray water should not be used inside the building. It is just for outside.

16. U.S. buildings account for what percentage of all potable water consumption?

   a. 10 percent

   b. 14 percent

   c. 15 percent

   d. 50 percent

17. Which standard sets the maximum flush and flow rates in buildings?

   a. EPAct of 1992

   b. Water Policy Act of 1992

   c. International Organization for Standardization 14044

   d. ASHRAE 55.5

18. A project owner hired a LEED consultant to determine the default gender ratio for full-time equivalent occupants. Which ratio did the LEED consultant advise using?

   a. 40:60 ratio

   b. 60:40 ratio

   c. 70:30 ratio

   d. 50:50 ratio

19. What percentage of all potable water consumption does the United States use?

   a. 10 percent

   b. 12 percent

   c. 14 percent

   d. 16 percent

20. What are the strategies for reducing indoor water use? Select two.

   a. Replace all the toilet fixtures with new ones

   b. Install efficient fixtures

   c. Reduce process water use

   d. Use reclaimed water in the shower

21. What is the benefit of WaterSense products?

    a. They are 5 percent more efficient than average products.

    b. They are 7 percent more efficient than average products.

    c. They are 15 percent more efficient than average products.

    d. They are 20 percent more efficient than average products.

22. Which water from toilets and urinals that is considered under all codes to be wastewater?

    a. Gray water

    b. Black water

    c. Cooling tower blowout

    d. Baseline water

23. Using the baseline water consumption, what is the flush rate for a public lavatory faucet?

    a. 1.6 gpf (6.0 Lpf)

    b. 0.5 gpf (1.9 Lpf)

    c. 1.6 gpm (1.9 Lpm)

    d. 0.5 gpm (1.9 Lpm)

24. Which two are examples of gray water?

    a. Wastewater from a shower

    b. Wastewater from a washing machine

    c. Reclaimed water from a cistern

    d. Water from toilets

25. What helps a building seeking a LEED certifications more in earning points for irrigation?

    a. Using potable water

    b. Using nonpotable water

    c. Using both a and b

    d. Using only gray water

26. Using waterless fixtures and faucets with aerators help to reduce?

    a. Outdoor water

    b. Indoor water

    c. Indoor and outdoor water

    d. Process water

27. A project team wants to calculate the baseline and design case for WE. What should the project team do?

    a. Use reclaimed water for all outdoor use

    b. Install an aerator at every indoor fixture

   c. Use water meters

   d. Use xeriscaping

28. In a LEED project, the owner was concerned with finding a way to track water usage trends, monitor fixture performance, and identify leaks. What did the LEED professional of the project advise doing?

   a. To use submetering

   b. To use WaterSense

   c. To use native and adapted species

   d. To increase the use of nonpotable water

29. Which water is used for domestic purposes such as cooking is?

   a. Nonpotable water

   b. Potable water

   c. Reclaimed water

   d. Gray water

30. A LEED professional is advising the project team members on the selection of kitchen faucet types. Because the LEED professional uses the baseline water consumption, so she advises them to use a kitchen faucet with which flush or flow rate?

   a. 0.5 gpm (1.9 Lpm)

   b. 2.2 gpm (8.3 Lpm)

   c. 1.6 gpf (6.0 Lpf)

   d. 2.5 gpm (9.5 Lpm)

31. How much reduction in water usage can be seen from installing submeters?

   a. 50 percent

   b. 60 percent to 70 percent

   c. 30 percent to 40 percent

   d. Around 50 percent

32. Boilers and cooling towers involve which type of water?

   a. Nonpotable water

   b. Process water

   c. Gray water

   d. Black water

33. What is the water usage of dual-flush water closets for liquid waste?

   a. 1.28–1.6 gpf

   b. 0.8–1.0 gpf

   c. 0.125 pgf

   d. 1.28 gpf

34. During the design phase, which of the following must be confirmed to avoid any conflict during construction?

    a. Submetering type

    b. Toilet fixtures

    c. Alternative water sources

    d. None of the above

35. Some elements increase outdoor water use. Which is one such element?

    a. Turfgrass

    b. Plant density

    c. Submetering

    d. Xeriscaping

36. An owner has complained about indoor water use at his facility. How can the owner reduce the water bill (i.e., reduce excessive consumption)?

    a. Replace all the bathroom fixtures, faucets, and water fountains

    b. Install submeters

    c. Install a drip irrigation system

    d. Install cisterns

37. Which results in the production of gray water?

    a. Washing machine

    b. Toilet

    c. Collected rainwater

    d. City water

38. What is treated water used as?

    a. Potable water and must be labeled to prevent unintentional use

    b. Nonpotable water and must be labeled to prevent unintentional use

    c. A source for irrigation only

    d. A black water

39. A project team is working with an owner to renovate her old apartment complex, and in this endeavor the team wants to reduce the water consumption as much as possible. What are some good suggestions?

    a. Install submeters

    b. Place low-flow aerators in all faucets

    c. Use only Energy Star appliances

    d. Install only new water heaters

40. How much more efficient is drip irrigation than traditional sprinklers?

    a. 10 percent

    b. 20 percent

    c. 80 percent

    d. 90 percent

41. An owner who would like to develop his backyard wants to eliminate the need for irrigation. How can he do that?

    a. Drip irrigation

    b. Xeriscaping

    c. Mulching

    d. Density plant to reserve more water

42. As a LEED professional, what type of water would you recommend for irrigation?

    a. Black water

    b. Municipal water

    c. Gray water

    d. Red water

43. As a LEED professional, how can you reduce indoor water use?

    a. Use a compositing toilet

    b. Use smaller toilet seats

    c. Use smaller kitchen sinks

    d. Install fixtures and faucets from reputable vendors

44. What is the benefit of providing a manufacturer's cut sheets for indoor water fixtures?

    a. This proves that a toilet fixture is qualified for water reduction.

    b. It is not required to provide the manufacturer's cut sheets.

    c. The manufacturer's cut sheets are only required for outdoor water fixtures.

    d. The manufacturer's cut sheets can only be provided for energy devices in indoor facilities.

45. Which reduction is highly prioritized over others by the sustainable community?

    a. Outdoor water use

    b. Indoor water use

    c. Cooling water use

    d. Reclaimed water use

46. Harvesting rainwater has numerous benefits. Which of the below is not an applicable choice?

    a. To reduce potable water use

    b. To reduce runoff

    c. To reduce outdoor water use

    d. To reduce landscape fertilization needs

47. Which approach is helpful in reducing water consumption?

    a. Permeable surface increase

    b. Landscape area increase

    c. Landscape that requires less irrigation

    d. Rainwater collected in cisterns

48. Rainwater is _____.

    a. Similar to municipality water

    b. Black water

    c. Gray water

    d. Nonpotable water

49. What entity or organization decides on the type of water used outdoors?

    a. City/county's code and regulations

    b. Project team members

    c. Owner

    d. Mechanical engineer

50. How can a team reduce potable water consumption?

    a. Install a leach field and septic tank system

    b. Use permeable surfaces in multiple areas

    c. Use toilets with a double-flush feature

    d. Use a flat commercial roof instead of a hip roof

51. The EPA recommends indoor products with which label?

    a. Full Time Equivalent Occupancy (FTE)

    b. EPAct of 1992

    c. WaterSense

    d. ASHRAE

52. To calculate the baseline daily water usage of a project, which of the following requires the toilets to use no more than 1.6 gallons of water per flush and all urinals to use no more than 1.0 gallon per flush?

    a. ASHRAE 99-2012

    b. ASHRAE 55

    c. EPAct of 1992

    d. Commissioning authority

53. The water use reduction prerequisite requires _____ reduction of the design case compared to the baseline case.

    a. 5 percent

    b. 10 percent

    c. 15 percent

    d. 20 percent

References

Allen, D. T., Shonnard, D. R., Huang, Y., & Schuster, D. (2016). Green Engineering Education in Chemical Engineering Curricula: A Quarter Century of Progress and Prospects for Future Transformations, ACS Publications. https://pubs.acs.org/doi/full/10.1021/acssuschemeng.6b01443?casa_token=ildzY6YKnb8AAAAA%3AEJviQ0NbPXiOuSr_mx95Vj4jv_30UT-lYV7shjuqSI1Jh1f6a7bp1__z1kdaMHbIMt4Q4Gm-hAZV9NRAz.

Alshareef, H. A., Clark, A., Milyard, A., Robert, H., & Hund, B. LEED Process Assessments and Efficiency Improvements for Renovated Buildings. *El Río: A Student Research Journal*, 3 (1): 14–21, 2020.

Commercial Buildings Energy Consumption Survey (CBECS), published by Energy Information Administration (EIA), (2018). Accessed 2020. http://eia.doe.gov/emeu/cbecs/background/html.

Maupin, M. A., Kenny, J. F., Hutson, S. S., Lovelace, J. K., Barber, N. L., & Linsey, K. S. (2014). *Estimated use of water in the United States in 2010* (No. 1405). US Geological Survey. Accessed October 2019. https://pubs.er.usgs.gov/publication/cir1405.

Knox, H. J., lll. (2007). Case study: Adobe's "Greenest Office in American" Sets the Bar for Corporate Environmentalism. U.S. Green Building Council. Accessed February 2021. https://www.fmlink.com/articles/case-study-adobes-greenest-office-in-america-sets-the-bar-for-corporate-environmentalism/.

Massachusetts Water Resources Authority. (2021). Water Efficiency and Management for Commercial Buildings. Accessed September 2022. https://www.mwra.com/comsupport/ici/commercialbuildings.htm.

Solley, W., Robert, B., Pierce, R., and Perlman, H. A. (1998). Estimated use of water in the U.S. in 1998. U.S. Geological Survey. https://pubs.usgs.gov/circ/1998/1200/report.pdf. Accessed September 2022.

U.S. Environmental Protection Agency (EPA). (2018). WaterSense. Why Water Efficiency? http://www.epa.gov.owm/water-efficincy/water/why/html.

U.S. Geological Survey. (2008). USGS Study Documents Water Levels Changes in High Plains Aquifer.

U.S. Green Building Council. (2021). LEED Certified Project List. https://www.usgbc.org/projects/. Accessed September 2022.

UN water, statistics: Graphs & Maps, http://www.unwater.org/statistics/en/ (Accessed July 2019). https://www.unwater.org/water-facts/.

TGBMIR (2009), by Consulting-Specifying engineer, The Green Building Market and Impact Report 2009 release: an annual green buildings report highlights the impact if LEED-certified buildings (Accessed January 2020). https://www.csemag.com/articles/the-green-building-market-and-impact-report-2009-is-released/.

Answers

1. **b** The water daily rate depends on the occupancy and fixture types.
2. **c & d**
3. **d**
4. **a**
5. **b**

6. c

7. c

8. a

9. b

10. b

11. d

12. b

13. a

14. d

15. b

16. b

17. a

18. d

19. c

20. b & c

21. d

22. b

23. d

24. a & b

25. b

26. b

27. c

28. a

29. b

30. b

31. c

32. b

33. b

34. c Reclaimed water must be identified and documented at an early stage of the project related to codes and regulations.

35. a Submetering is for monitoring water use, and xeriscaping is for outdoor water reduction. The density of plants helps to retain water and prevent runoff. The correct answer is a, which is turfgrass. Turfgrass requires a lot of irrigation.

36. c Submeters help to monitor the usage so users can adjust their consumption accordingly. Option a is too expensive, and the other options are for the outdoor environment.

37. a

38. b

39. **a & b** Installing new water heaters may not necessarily help reduce water; the same is true for Energy Star appliances. The first two options are the best.

40. **d**

41. **b** The question asks how the owner can eliminate irrigation and not reduce its usage.

42. **c** Black water is not recommended for irrigation, and municipal water is a potable water, for which it is recommend to reduce its use as much as possible. There is no red water.

43. **a** Smaller toilet seats and kitchens will not affect water use, and neither will reputable vendors.

44. **a** One of the documentation processes for water-saving fixtures is to include the manufacturer's cut sheets to demonstrate the efficiency of the fixture.

45. **b** The Indoor Water Use Reduction credit (from the scorecard) has six reward points, which is the highest among all the other WE category's credits.

46. **d**

47. **c** Option d is for rainwater management. Option a is not applicable, and option b will increase water use.

48. **d**

49. **a**

50. **c** A leach field does not save potable water. All other options are not applicable, and option c is correct.

51. **c**

52. **c** The EPAct of 1992 sets the maximum flush and flow rates in buildings.

53. **d**

CHAPTER 7

Energy and Atmosphere (EA)

The Energy and Atmosphere (EA) category approaches energy from a holistic perspective, addressing energy use reduction, energy-efficiency design strategies, and renewable energy sources.

The current worldwide mix of energy resources is weighted heavily toward oil, coal, and natural gas, according to statistical data from the International Energy Agency (CETI, 2022). In addition to emitting greenhouse gases, these resources are nonrenewable, their quantities are limited, or they cannot be replaced as fast as they are consumed. Though estimates regarding the remaining quantity of these resources vary, it is clear that the current reliance on nonrenewable energy sources is not sustainable and involves increasingly destructive extraction processes, increasing supplies, escalating market prices, and national security vulnerability. According to the U.N. Environment Program, buildings are significant contributors to these problems, accounting for approximately 40 percent of the total energy used today.

Energy efficiency in a green building starts with a focus on a design that reduces overall energy needs, such as building orientation and glazing selection and the choice of climate-appropriate building materials. Strategies such as passive heating and cooling, natural ventilation, and high-efficiency heating, ventilating, and air-conditioning (HVAC) systems partnered with smart controls further reduce a building's energy use. The generation of renewable energy on the project site or the purchase of green power allows portions of the remaining energy consumption to be met with non-fossil-fuel energy, lowering the demand for traditional sources.

The commissioning process is critical for ensuring high-performance buildings. Early involvement of a commissioning authority (CxA) helps prevent long-term maintenance issues and wasted energy by verifying that the design meets the owner's project requirements (OPRs) and functions as needed. In an operationally effective and efficient building, the staff understands what systems are installed and how they function. Staff must have training and be receptive to learning new methods for optimizing system performance so that efficient design is carried through to efficient performance.

The EA category recognizes that the reduction of fossil fuel use extends far beyond the walls of the building. Projects can contribute to increasing the electricity grid's efficiency by controlling in a demand response program. Demand response allows utilities to call on buildings to decrease their electricity use during peak times, reducing the strain on the grid and the need to operate more power plants, thus potentially avoiding the costs of constructing new plants. Similarly, on-site renewable energy not only moves

77

the market away from dependence on fossil fuels but also may be a dependable local electricity source that avoids transmission losses and strain on the grid.

The American Physical Society has found that if current and emerging cost-effective energy efficiency measures are employed in new buildings and in existing buildings as their heating, cooling, lighting, and other equipment is replaced, the growth in energy demand from the building sector could fall from a projected 30 percent increase to zero between now and 2030 (APS report, 2008). The EA category supports the goal of reduced energy demand through credits related to reducing usage, designing for efficiency, and supplementing the energy supply with renewables.

Buildings consume approximately 39 percent of the energy and 74 percent of the electricity produced annually in the United States according to the U.S. Department of Energy (Building Energy Data Book 2011). Generating electricity from fossil fuels such as oil, natural gas, and coal negatively affects the environment at each step of production and use, beginning with extraction and transportation, followed by refining and distribution, and ending with consumption. For example, coal mining disrupts natural habitats and can devastate landscapes. Acidic mine drainage degrades regional ecosystems. Coal is rinsed with water, producing billions of gallons of sludge that must be stored in ponds. Mining itself is a dangerous occupation in which accidents and the long-term effects of breathing coal dust can shorten the life span of coal miners.

Electricity is most often generated by burning fossil fuels, whose combustion releases carbon dioxide and other greenhouse gases that contribute to climate change. Coal-fired plants accounted for more than half of U.S. electricity generation in 2006 according to the U.S. Department of Energy, Office of Energy Efficiency and Renewable Energy (Advanced Manufacturing Office workshop report, 2021). Burning coal releases harmful pollutants such as carbon dioxide, sulfur dioxide, nitrogen oxides, small particulates, and mercury. Each megawatt of coal-generated electricity releases into the atmosphere an average of 2249 pounds of carbon dioxide, 13 pounds of sulfur dioxide, and 6 pounds of nitrogen oxides. More than 65 percent of the sulfur dioxide released into the air, or more than 13 million tons per year, comes from electricity generation, primarily by coal-burning generators. Mining, processing, and transporting coal to power plants create additional emissions, including methane vented from the coal during transport.

Natural gas, nuclear fission, and hydroelectric generators all have adverse environmental consequences as well. Natural gas is a major source of nitrogen oxide and greenhouse gas emissions. Nuclear power increases the potential for catastrophic accidents and raises significant waste transportation and disposal issues. Hydroelectric-generating plants disrupt natural water flows and disturb aquatic habitats.

Green building addresses those issues in two ways. First, it reduces the amount of energy required for building operations; second, it uses more benign forms of energy. The better the energy performance of a building, the fewer greenhouse gases are emitted from energy production. Electricity generation using sources other than fossil fuel also reduces the environmental impacts from a building's energy use. Additionally, improved energy performance results in lower operating costs. As global competition for fuel accelerates, the rate of return on energy efficiency measures improves.

Energy Performance

The energy performance of a building depends on its design. Its massing and orientation, materials, construction methods, building envelope, and water efficiency as well as the HVAC and lighting systems determine how efficiently the building uses energy.

Therefore, the most effective way to optimize energy is to use an integrated, whole-building approach. Collaboration among all team members, beginning at project inspection, is necessary in designing building systems.

Green schools prove a unique educational opportunity. Exploring the interdependence of building systems, design, and energy performance can foster ecological literacy. This study also helps students learn to think systemically about complex, real-world issues by seeing the relationships among buildings, human activities, resources consumption, and environmental impacts (Alshareef et al., 2020). If real-time monitoring is incorporated into the school's energy system, students can see how the building responds to sunlight, moisture, temperature, and other environmental conditions, which can support the study of thermodynamics, earth science, and other subjects.

Tracking Building Energy Performance: Designing, Commissioning, Monitoring

Projects that achieve any level of LEED (Leadership in Energy and Environmental Design) certification must at a minimum perform better than the average building. Specific levels of achievement beyond the minimum are awarded a proportional number of points. First, the building must be designed to operate at a high performance level. Next, it must be commissioned to ensure that what has been constructed meets the design intent.

Performance of the Building's Energy Systems

The design of new facilities must be based on the designated mandatory and prescriptive requirements of the American Society of Heating, Refrigeration and Air-Conditioning Engineers (ASHRAE) 90.1-2010 or local code approved by the U.S. Green Building Council (USGBC), whichever is more stringent. In addition, optimization of building energy performance beyond ASHRAE 90.1-2010 is required in EA prerequisite 2, Minimum Energy Performance. This can be accomplished through building energy simulation modeling or prescriptive options.

Commissioning begins with the development of the OPRs, followed at a minimum by additional steps that include creation of a formal commissioning plan, verification of equipment installation, and submission of a final report. Enhanced commissioning includes additional tasks, such as design and contract submittal reviews, creation of a formal systems manual, verification of staff training, and a follow-up review before the warranty period ends.

Commissioning optimizes energy and water efficiency by ensuring that systems are operating as intended, thereby reducing the environmental impacts associated with energy and water usage. Additionally, commissioning can help ensure that indoor environmental quality is properly maintained. Properly executed commissioning can substantially reduce costs of maintenance, repairs, and resource consumption, and higher indoor environmental quality can enhance occupants' productivity.

Monitoring the performance of building systems begins with establishing a measurement and verification plan based on the best practices developed by the International Performance Measurement and Verification Protocol. The plan must cover at least 1 year of postconstruction occupancy. Monitoring involves appropriate measuring instruments and can include the energy modeling.

Managing Refrigerants to Eliminate Chlorofluorocarbons

The release of chlorofluorocarbons (CFCs) from refrigeration equipment destroys ozone molecules in the stratosphere through a catalytic process and harms Earth's natural shield from incoming ultraviolet radiation. CFCs in the stratosphere also absorb infrared radiation and create chlorine, a potent greenhouse gas. Banning the use of CFCs in refrigerants slows the depletion of the ozone layer and mitigates climate change.

The standard practice for new buildings is to install equipment that does not use CFC-based refrigerants. In LEED, points are awarded for systems that use refrigerants with a low potential for causing ozone depletion and climate change.

Using Renewable Energy

Teams have two opportunities to integrate renewable energy strategies into the LEED project: using on-site renewable energy systems and buying off-site green power. Projects can have integrated systems that incorporate on-site electrical (photovoltaic, wind, wave, tidal, and biofuel-based); geothermal (deep-earth water or steam); or solar thermal (including collection and storage components) power. Credit for off-site renewable green power is achieved by contracting for a minimum purchase of green power.

Energy generation from renewable sources such as solar, wind, and biomass avoids air and water pollution and other environmental consequences associated with producing and consuming offal and nuclear fuels. Although hydropower is considered renewable, it can have harmful environmental effects, such as degrading water quality, altering fish and bird habitats, and endangering species. Low-impact hydropower, if available, is recommended.

Renewable energy minimizes acid rain, smog, climate change, and human health problems resulting from air contaminants. In addition, using renewable resources avoids the consumption of fossil fuels, the production of nuclear waste, and the environmentally damaging operation of hydropower dams.

Renewable alternatives may be less expensive than traditional power in some areas. Utility and public benefit fund rebates may be available to reduce the initial cost of purchasing and installing renewable energy equipment. In some states, net metering can offset on-site renewable energy costs when excess electricity generated on-site is sold back to the utility.

Benefits and Issues to Consider

Benefits of commissioning include reduced energy use, lower operating costs, fewer contractor callbacks, better building documentation, improved occupant productivity, and verification that the system performs in accordance with the OPRs.

Environmental Issues

Facilities that do not perform as intended may consume significantly more resources over their lifetimes than they should. Commissioning can minimize the negative impacts buildings have on the environment by helping verify that buildings are designed and constructed to operate as intended and in accordance with the OPRs.

Economic Issues

If commissioning has not been previously included as part of the project delivery process, the costs associated with commissioning may be met with initial resistance. When the long-term benefits are taken into consideration, however, commissioning can be seen as a cost-effective way to ensure that the building is functioning as designed and that planned energy savings are realized in the operation of the building.

Improved occupant well-being and productivity are other potential benefits when building systems function as intended. Proper commissioning of building systems can reduce employee illness, tenant turnover and vacancy, and liability related to indoor air quality, and it can avoid premature equipment replacement.

Implementation

For LEED design and construction projects, the scope of services for the CxA and project team should be based on the OPRs. The commissioning process activities must address the commissioned systems noted in the EA Prerequisite 1 requirements. Other systems, such as the building envelope, stormwater management systems, water treatment systems, and information technology systems, may also be included in the commissioning process at the owner's discretion. EA Credit 3 requires the CxA to be involved early in the process to help facilitate a commissioning design review and a commissioning documentation review. As the project nears completion, enhanced commissioning requires oversight of staff training, a walk-through 10 months after completion, and the completion of a system manual.

Timeline and Team

The commissioning process is a planned, systematic quality control process that involves the owner, users, occupants, operations and maintenance staff, design professionals, and contractors. It is most effective when begun at project inception. All members of the project team are encouraged to participate in the commissioning activities as part of a larger commissioning team. The team approach to commissioning can speed the process and add a system of checks and balances.

Keywords and Definitions

Biofuel-Based System: Power systems that run on renewable fuels derived from organic materials such as wood by-products and agricultural waste. Examples of biofuels include untreated wood waste, agricultural crops and residues, animal waste, other organic waste, and landfill gas.

Biomass: Plant material from trees, grasses, or crops that can be converted to heat energy to produce electricity.

Geothermal Heating Systems: Use of pipes to transfer heat from underground steam or hot water from heating, cooling, and hot water reservoir. The system retrieves heat during cool months and returns heat in summer months.

Green Power: Synonymous with renewable energy.

Hydropower: Electricity produced from the downhill flow of water from rivers or lakes.

Photovoltaics (PV): Electricity from photovoltaic cells that convert the energy in sunlight into electricity.

Renewable Energy: Comes from sources that are not depleted by use. Examples include energy from the sun, wind, and small (low-impact) hydropower, plus geothermal energy and wave and tidal systems. Ways to capture energy from the sun include photovoltaic, solar thermal, and bioenergy systems based on wood waste, agricultural crops or residue, animal and other organic waste, or landfill gas.

Wind Energy: Electricity generated by wind turbines.

Questions

The questions and multiple-choice answers that follow were designed to provide a deeper insight and look into the EA category.

1. The credits in the EA category goals are to reduce energy demand, increase energy efficiency, and replace fossil fuel use with renewable energy and carbon offsets. What else could fall under these goals?

   a. Eliminate the use of harmful refrigerants

   b. Reduce low-emitting materials

   c. Reduction of total waste materials

   d. Increase daylighting

2. Energy baseline consumption is established by third-party standards organizations. Which origination below is NOT included?

   a. American Society of Heating, Refrigeration and Air-Conditioning (ASHRAE)

   b. American National Standards Institute (ANSI)

   c. Illuminating Engineering Society of North America (IESNA)

   d. Energy Star Portfolio Manager

3. What does submetering allow facility managers to do?

   a. Track energy costs and usage by area

   b. Reduce energy demand

   c. Look for alternatives

   d. Incorporate passive strategies

4. What is the most widely used building benchmarking system in the United States?

   a. ASHRAE 90.1

   b. ASHRAE 62.2

   c. Energy Star Portfolio Manager

   d. Energy Policy Act of 1992

5. What is the label Energy Star certified label used for?

   a. Showerheads

   b. Refrigerators

    c. Irrigation systems

    d. Kitchen faucets

    e. Passive solar architectural features

6. What are renewable energy certificates (RECs) another form of?

    a. Green-e certification program

    b. Green power

    c. Solar power

    d. Energy Star label

7. What is the basis from which all design, construction, acceptance, and operational decisions are made?

    a. Commissioning (Cx)

    b. Commissioning authority (CxA)

    c. Owner's project requirements (OPRs)

    d. Demand Responses

8. While passive solar design helps with heating, what does passive ventilation assist with?

    a. Daylight

    b. Lighting pollution

    c. Cooling

    d. On-site renewable energy

9. Which program helps to alleviate the need to build more energy infrastructure to meet the needs of every user when demand is high and supply is low?

    a. Natural ventilation

    b. Mechanical ventilation

    c. Demand response

    d. High-efficiency appliances

10. EA Credit: Enhanced Refrigerants Management attempts to mitigate the tradeoff by allowing only the use of refrigerants that are naturally occurring or synthetic with an Ozone Depletion Potential (OPD) of ____ and a Global Warming Potential (GWP) of less than ____.

    a. 0, 50, respectively

    b. 50, 0, respectively

    c. 50, 50, respectively

    d. 100, 50, respectively

11. The airflow of the building is influenced by

    a. The shape of the building

    b. The building footprint

    c. The region where the building is located

    d. Time of year (winter vs. summer)

12. EA Credit: Green Power and Carbon Offsets awards 1 to 2 points for engaging in a green power contract with a utility for a minimum of what term?

    a. 3 years

    b. 4 years

    c. 5 years

    d. 6 years

13. A project team is working on qualifying their project for points under EA Credit: Green Power and Carbon Offsets. Which program are the qualifications based on?

    a. WaterSense water efficiency

    b. Green-e certification program

    c. Department of Energy program

    d. Retrocommissioning program

14. By which year does the Montreal Protocol ban CFCs and phase out hydrochlorofluorocarbons (HCFCs)?

    a. 2025

    b. 2029

    c. 2030

    d. 2050

15. Which of the following is reduced by using carbon offsets?

    a. Greenhouse gas emission

    b. HCFCs

    c. CFCs

    d. ASHRAE

16. Which commissioning process can be performed on existing buildings to identify and recognize system improvements that make the building more suitable for current use?

    a. Commissioning authority CxA)

    b. Commissioning

    c. Retrocommissioning

    d. None of the above

17. In commercial buildings, on average which product's end use consumes the most energy?

    a. Dishwasher

    b. Space heater

    c. Refrigerator

    d. A TV running all day

18. What year does the Montreal Protocol phase out HCFCs?
    a. 2025
    b. 2030
    c. 2050
    d. HCFCs will never phase out

19. What does ozone depletion potential contain?
    a. HCFCs
    b. CFCs
    c. ODP
    d. GWP

20. When does commissioning take place?
    a. Before final inspection
    b. During substantial completion
    c. Before the project starts
    d. After the final inspection

21. Which refrigerants are considered environmentally benign? Select two.
    a. Hydrocarbons
    b. Propane
    c. Coal
    d. Crude oil

22. What influences the amount of light allowed in a building?
    a. Building shape
    b. Building orientation
    c. Building materials
    d. Building footprint

23. The goal of energy efficiency is to
    a. meet the demand on the energy use
    b. provide demand response
    c. accomplish the same amount of work with less energy expended
    d. balance the supply and demand

24. What are the two largest renewable energy sources?
    a. Solar power
    b. Wind
    c. Biomass
    d. Hydropower

25. Which is an example of a plug load?

    a. Cooling system

    b. Heating system

    c. Kitchen light

    d. Computer

26. Which encourages large energy users to reduce loads during peak use time in exchange for reducing cost?

    a. Green power

    b. Carbon offsets

    c. Demand response

    d. Enhanced commissioning

27. A project uses a required amount of energy from the grid and no more than it can produce on-site. What is an equal input to output ratio called when using energy?

    a. Advanced energy metering

    b. Net-zero energy

    c. Green power

    d. Thermal comfort

28. What are the three things an owner can do to ensure energy performance is working sufficiently?

    a. Attend construction meetings

    b. Attend weekly charette

    c. Provide training and incentives for facility managers and occupants

    d. Provide yearly maintenance and tune-up on all systems

29. Which is in charge of creating and developing the lighting industry standards?

    a. Energy use intensity (EUI)

    b. Energy Star Portfolio Manager

    c. IESNA

    d. Lighting power density (LPD)

30. Which is a standard that sets the minimum requirements for energy-efficient design of most buildings, except low-rise residential buildings?

    a. ASHRAE 62.1-2010

    b. ASHRAE 90.1-2010

    c. ASHRAE 55-2010

    d. ASHRAE 55.1-2010

31. Which is a renewable energy that is purchased from a utility?

    a. Green-e certification program

    b. Carbon offsets

    c. Green power

    d. Energy modeling

32. What is the friendliest refrigerant system to date?

    a. CFCs

    b. HCFCs

    c. Hydrofluorocarbons (HFCs)

    d. HCFCs and HFCs

33. Select three natural refrigerants.

    a. Chlorofluorocarbons

    b. Carbon dioxide

    c. Ammonia

    d. Propane

34. A project team is brainstorming ideas for reducing the energy demand on a new building that is going to be built soon. What will help the team in their endeavor?

    a. Increasing the square footage of the building to allow more airways

    b. Increasing the window ratio to the floor without using shading devices

    c. Participating in a demand response program

    d. Selecting an optional building orientation

35. Which is a third party who reports directly to the owner and oversees documents and system assemblies, including planning, designing, installing, and testing?

    a. Retrocommissioner

    b. CxA

    c. Project administrator

    d. LEED online specialist

36. According to the Montreal Protocol (Montreal amendment), developing countries that started phasing out HCFCs since 2013 agreed to phase HCFCs completely by which year?

    a. 2025

    b. 2027

    c. 2030

    d. 2035

37. A design team is discussing an initial building layout for a healthcare facility, so the team's goal, besides creating a sustainable building, is to gain as many reward points as possible for a higher level certification. Which statement is true?

    a. Enhance space categorization.

    b. Maintain small module spacing.

c. Design one giant open space.

d. Healthcare facilities are difficult to gain reward points with.

38. A building's energy use works as a function of its size. What does this refer to?

    a. Single light task

    b. Daylight

    c. Energy use intensity

    d. Fossil fuels

39. Which project is eligible for LEED certification?

    a. New building that does not use refrigerants

    b. Old building that uses CFC refrigerants

    c. A new building that uses CFC refrigerants

    d. An old building that uses natural ventilation and CFC refrigerants

40. What is an example of occupied space?

    a. Mechanical room

    b. Data center floor area

    c. Closet space

    d. Gymnasium

41. Which is a purchasable form of trade that funds projects that reduce greenhouse gas emissions, such as forest restoration, power plant and factory updates, or increase to the energy efficiency of buildings and transportation?

    a. REC

    b. Carbon offsets

    c. Green power

    d. Photovoltaic energy

42. Which actively reduces greenhouse gas emissions?

    a. Using refrigerants that contain CFCs

    b. Including ozone depletion potential

    c. Using carbon offsets

    d. Using an energy start portfolio manager

43. Select two examples of process energy devices or equipment.

    a. Electric stove

    b. Escalator

    c. Central heating and cooling system

    d. Tankless water heater

44. Select two examples of nonprocess energy devices or equipment.

    a. Microwave

    b. Space heater

    c. Tankless water heater

    d. Sump pump

45. Select all applicable forms of renewable energy on-site.

    a. Carbon offsets

    b. RECs

    c. Animal waste

    d. Agricultural crops or waste

46. Refrigerants' harms to the environment are measured and evaluated based on what value?

    a. Usage by tons

    b. ODP and GWP

    c. Manufacturer's reputation and reliability

    d. Mechanical ventilation

47. According to the USGBC, the baseline performance is the annual energy cost for a building design, which is used as a baseline for comparison with above-standard design. How is the baseline energy performance calculated?

    a. Using ASHRAE 55-2010

    b. Using ASHRAE 90.1

    c. Using IESNA

    d. Using performance monitoring

48. What do solar panels installed on site benefit?

    a. Reduced light pollution

    b. Reduced high-reflectance materials

    c. Reduced greenhouse gas emissions

    d. Reduced impervious hardscape

49. To purchase an REC, what is the minimum length of a contract signed with the third party?

    a. 1 year

    b. 2 years

    c. 15 years

    d. 5 years

50. An existing building's owner would like to consult with a LEED professional to identify and recognize system improvements that would make the building more suitable for its current use. Which is the best advice?

    a. Use retrocommissioning

    b. Use passive energy

    c. Redo the building envelop

    d. Install waterless urinals

51. After occupancy, how long is commissioning warranted?

    a. 3 years

    b. 2 years

    c. 1 year

    d. 5 years

52. What does retrocommissioning help with?

    a. Selecting a new building envelop

    b. Providing energy efficiency and improvement

    c. Finding building layout configuration

    d. Choosing a building orientation

53. What is the essential purpose of commissioning?

    a. To meet city/county codes and requirements

    b. To satisfy the demand for commissioning, which may vary from project to project and from owner to another

    c. To meet the OPRs

    d. To fulfill the demand respond program's requirements whenever needed

54. What is the unit measure of LPD?

    a. The installed lighting power per unit area

    b. The installed lighting power per unit volume

    c. The megawatt

    d. The British thermal unit (Btu)

55. Which is in charge of banning harmful refrigerants?

    a. EPA

    b. ASHRAE 55

    c. Montreal Protocol

    d. ANSI

56. An owner wants to build a zero-net energy house in the countryside where there is no utility service. Which is the best practice for that?

    a. Install photovoltaic system

    b. Purchase renewable energy credits

    c. Purchase carbon offset

    d. Harvest rainwater for daily use

References

Alshareef, H. A., Vallejos, K., Aguilar, P., Leasure, D., & Trueblood, C. (2020). Involving Stakeholders at early Stage of the Design Process to Improve Reward Points Allocation. *El Río: A Student Research Journal, 3*(1), 3–13.

Clean Energy Transition Indicators (CETI), International Energy Agency (IEA) (2022). Accessed September 2022. https://www.iea.org/data-and-statistics/data-tools/clean-energy-transition-indicators.

Energy Efficiency Crucial to Achieving Energy Security and Reducing Global Warming, States Leading Scientists Report, American Physical Society (APS), September 16, 2008. Accessed August 2022. https://www.aps.org/newsroom/pressreleases/energyreport.cfm.

Building Energy Data Book (2011), Department of Energy. Accessed September 2022. https://ieer.org/wp/wp-content/uploads/2012/03/DOE-2011-Buildings-Energy-DataBook-BEDB.pdf.

Department of Energy, Office of Energy Efficiency and Renewable Energy: Advanced Manufacturing Office, (2021). Workshop on integrated sensor systems for manufacturing applications. Workshop report. Accessed January 2022. https://www.energy.gov/sites/default/files/2021-07/AMO%20Semiconductor%20Workshop%20I%20Integrated%20Sensor%20Systems%20Report.pdf.

Answers

1. **a**   The credits in the EA category have five goals. Besides the ones mentioned in the problem statement and option a, the last one is to monitor ongoing performance.

2. **d**   The Energy Star Portfolio Manager is an EPA provided building benchmarking system that is the most widely used in the United States. Options a, b, and c are all third-party standards organizations that establish energy baselines.

3. **a**   Submeters do not reduce energy demand or provide other benefits except tracking energy usage, so users can monitor their consumption and also measure consumption cost.

4. **c**   The Energy Star Portfolio Manager is an EPA-provided building benchmarking system that is the most widely used in the United States. The energy and water date for a building is entered into a web-based tool, which then displays where the building type falls in whole-building energy use compared to other buildings of the same type.

5. **b**   The Energy Star rating, which not every appliance has, means that a product meets certain federally mandated guidelines regarding energy efficiency. Certified products must deliver the features and performance demanded by consumers in addition to increased energy efficiency. Option b is the only appliance among the others that answers the question.

6. **b**

7. **c**

8. **c**

9. **c**

10. **a**

11. **a**   It is true that all the options available influence airflow, but option a is the most influential among all the others. Without the shape and configuration of the building, it will be hard for airflow to enter the building.

12. **c**

13. **b**
14. **c**
15. **a**
16. **c**
17. **b**
18. **b**
19. **b**
20. **b**
21. **a & b**
22. **b**
23. **c**
24. **c & d**
25. **d**
26. **c**
27. **b**
28. **c & d**
29. **c**
30. **b**
31. **c**
32. **c**
33. **b, c, d**
34. **d**   A demand response program does not reduce the energy demand, and the same holds for the window-to-floor ratio. Sustainable buildings encourage and support reducing square footage, not increasing it. Building orientation influences the energy demand and reduces it for the life span of the building.
35. **b**   A LEED online specialist does not have that level of authority, and option a is irrelevant. Option b could have that level of responsibility if he or she is also a commissioning authority.
36. **c**   The developed countries have phased out HCFCs since 2020, but the developing countries are phasing out HCFCs by 2030.
37. **a**   Depending on the space categorization, point allocations may vary. Some credit requirements may be applicable to certain spaces, while others may not be. For example, the use of absorptive ceiling tiles can improve the acoustic performance of a building, but in other areas, they might not be enough and sound reflection and sound masking are more applicable.
38. **c**   Energy use intensity is a measure of the building's size per unit of floor space.
39. **a**   The most preferred approach is using no refrigerant when possible. Chloro-fluorocarbon (CFCs) refrigerant are banned.
40. **d**   Check out the requirements for occupied and unoccupied spaces.

41. **b** The problem statement is the definition of carbon offsets.

42. **d** Check out the definition of carbon offsets.

43. **a & b** All process energy equipment has a plug load.

44. **c & d** Nonprocess equipment do not have a plug load, but have built-in components.

45. **c & d** Options c and d fall under biofuel materials.

46. **b** ODP and GWP are the relative amount of degradation to the environment.

47. **b**

48. **c** Reducing greenhouse gas emission is a way to reduce using nonrenewable energy, such as fossil fuel.

49. **d**

50. **a** Check out the definition of retrocommissioning.

51. **c** The commissioning process begins during the predesign phase and continues to closeout. The warranty of commissioning is 1 year after construction.

52. **b** Since retrocommissioning is mainly used on existing buildings, options a, c, and d are inapplicable because these options are for new buildings.

53. **c**

54. **a**

55. **c**

56. **a** On-site renewable energy is the best practice. Rainwater is not applicable here because the owner wants to achieve zero-net energy. The other options are not reasonable even if they are doable.

CHAPTER 8

Materials and Resources (MR)

The Materials and Resources (MR) credit category focuses on minimizing the embodied energy and other impacts associated with the extraction, processing, transport, maintenance, and disposal of building materials. The requirements are designed to support a life-cycle approach that improves performance and promotes resource efficiency. Each requirement identifies a specific action that fits into the larger context of a life-cycle approach to embodied impact reduction.

Building operations generate a large amount of waste on a daily basis. Meeting the Leadership in Energy and Environmental Design (LEED) MR credits can reduce the quantity of waste while improving the building environment through responsible waste management and materials selection. The credits in this section focus on two main issues: the environmental impact of materials brought into the project building and the minimization of landfill and incinerator disposal for materials that leave the project building.

The Waste Hierarchy

Construction and demolition waste constitutes about 40 percent of the total solid waste stream in the United States according to the U.S. Environmental Protection Agency (EPA, RCR 2012) and about 25 percent of the total waste stream in the European Union according to the European Commission Service Contract on Management of Construction and Demolition Waste report (DGXI, European Commission: final report 1999). In its solid waste management hierarchy, the EPA ranks source reduction, reuse, recycling, and waste to energy as the four preferred strategies for reducing waste. The MR section directly addresses each of these recommended strategies.

Source reduction appears at the top of the hierarchy because it avoids environmental harms throughout a material's life cycle, from supply chain and use to recycling and waste disposal. Source reduction encourages the use of innovative construction strategies such as prefabrication and designing to dimensional construction materials, thereby minimizing material cutoffs and inefficiencies.

Building materials reuse is the next most effective strategy because reusing existing materials avoids the environmental burden of the manufacturing process. Replacing existing materials with new ones would entail production and transportation of new materials, and it would take many years to offset the associated greenhouse gases

95

through increased efficiency of the building. LEED has consistently rewarded the reuse of materials.

Recycling is the most common way to divert waste from landfills. In conventional practice, most waste is sent to a landfill—an increasingly unsustainable solution. In urban areas, landfill space is reaching capacity, requiring the conversion of more land elsewhere and raising the transportation costs of waste. Innovations in recycling technology improve sorting and processing to supply raw materials to secondary markets, keeping those materials in the production stream longer.

Because secondary markets do not exist for every material, however, the next most beneficial use of waste material is conversion to energy. Many countries are lessening the burden on landfills through a waste-to-energy solution. In countries such as Sweden and Saudi Arabia, waste-to-energy facilities are far more common than landfills. When strict air quality control measures are enforced, waste to energy can be a viable alternative to extracting fossil fuels to produce energy.

In aggregate, LEED projects are responsible for diverting more than 80 million tons of waste from landfills, and this volume is expected to grow to 540 million tons by 2030 based on U.S. Green Building Council (USGBC, green building facts) green building facts. From 2000 to 2011, LEED projects in Seattle, Washington, diverted an average of 90 percent of their construction waste from the landfill, resulting in 175,000 tons of waste diverted (City of Seattle, LEED projects analysis). If all newly constructed buildings achieved the 90 percent diversion rate demonstrated by Seattle's 102 LEED projects, the result would be staggering. Construction is no longer waste; it is a resource.

Selecting Sustainable Materials

Material selection plays a significant role in sustainable building operations. During the life cycle of material, its extraction, processing, transportation, use, and disposal can have negative health and environmental consequences, polluting water and air, destroying native habitats, and depleting natural resources. Environmentally responsible procurement policies can significantly reduce these impacts. Consider the relative environmental, social, and health benefits of the available choices when purchasing materials and suppliers. For example, the purchase of products containing recycled content expands markets for recycled materials, slows the consumption of raw materials, and reduces the amount of waste entering landfills. Use of materials from local sources supports local economies while reducing transportation impacts.

Practicing Waste Reduction

Maintaining occupancy rates in existing buildings reduces redundant development and the associated environmental impact of producing and delivering new materials. Construction waste disposal through landfilling it or incineration contributes significantly to the negative environmental impacts of a building. Construction and demolition waste constitute about 40 percent of the total solid waste stream in the United States (Wasmi, 2016). In its solid waste management hierarchy, the U.S. (EPA, April 2022) ranks source reduction, reuse, and recycling as the three preferred strategies to reduce waste. Source reduction appears at the top of EPA's hierarchy because it minimizes environmental impacts throughout the material's life cycle, from the supply chain and use to recycling and waste disposal. Reuse of materials is ranked second because fewer materials are diverted from the waste stream and substituted for other materials with greater environmental impacts. Recycling does not have all the same benefits as source reduction and reuse, but it diverts waste from landfills and incinerators and lessens the demand for virgin materials.

Reducing Waste at Its Source

Source reduction, which includes reducing the overall demand for products, is the most economical way to reduce waste. In 2006 and according to the EPA solid waste and solid waste management, U.S. residents, businesses, and institutions produced more than 251 million tons of solid waste, a 65 percent increase since 1980 (EPA, April 2022). The amount is roughly equivalent to 4.6 pounds per person per day, a 25 percent increase since 1980. In addition, 7.6 billion tons of industrial solid waste are generated each year. Waste generation raises building costs in two ways. First, unnecessary materials (e.g., packaging) add to the cost of products purchased, and second, fees for waste collection and disposal rise as the amount of waste increases. Reducing the amount of waste is an important component of sustainable construction practices. A construction waste management plan is the first step in managing construction waste because it requires contractors to establish a system for tracking waste generation and disposal during construction.

Reusing and Recycling

Reuse of existing buildings, versus building new structures, is one of the most effective strategies for minimizing environmental impacts. By reusing existing building components, waste can be reduced and diverted from landfills. Reuse of an existing building results in less habitat disturbance and typically less new infrastructure such as utilities and roads. An effective way to reuse interior components is to specify them in construction documents. By reusing or recycling these materials, an increasing number of public and private waste management operations have reduced the volume of construction debris. Recovery typically begins on the job site with separation of debris into bins or disposal areas. Some regions have access to mixed waste processing facilities. The EPA reported (EPA, RCR 2012) that in 2007 there were 34 mixed waste processing facilities in the United States handling about 43 million tons of waste per day.

When selecting materials, it is important to evaluate new and alternative sources. Salvaged material can be substituted for new material, saving costs and adding character to the building. Recycled content materials reuse waste that would otherwise be disposed of in landfills or incinerators. Use of local materials supports the local economy and reduces transportation impacts. Using rapidly renewable materials may minimize natural resource consumption with the harvest cycle of the resources potentially matching the life of the material in buildings. Use of third-party certified wood improves the stewardship of forests and related ecosystems.

Recycling construction, demolition, and land-clearing debris reduces demand for virgin resources and has the potential to lessen the environmental and health burdens associated with resource extraction, processing, and transportation. Debris recycling also reduces dependence on landfills, which may contaminate groundwater and encroach on valuable open space. In addition, it lessens disposal in incinerators, which may contaminate groundwater and pollute the air. Effective construction waste management can extend the life of existing landfills, which in turn reduces the need for expansion or development of new landfills.

Over the past few decades, recycling has increased in the United States. In 1960, only 6.4 percent of U.S. waste was recycled; the amount climbed to 32.5 percent. Curbside recycling is now standard in many communities, and recycling facilities are available throughout the nation. In addition, many businesses, nonprofit organizations, and manufacturers have successful recycling programs that divert a wide range of materials from the waste stream.

Recycling provides materials for new products that would otherwise be manufactured from virgin materials. It avoids the extraction of raw materials and preserves landfill space. Recycling certain products, such as batteries and fluorescent light bulbs, prevents toxic materials from polluting the air and groundwater.

Reuse and recycling can also save money. Effective waste management benefits organizations by reducing the cost of waste disposal and generating revenue from recycling or resale proceeds.

Life-Cycle Assessment in LEED

Through credits in the MR category, LEED has instigated transformation of building products by creating a cycle of consumer demand and industry delivery of environmentally preferable products. LEED project teams have created demand for increasingly sustainable products, and suppliers, designers, and manufacturers are responding. From responsible harvested wood to increased recycled content to bio-based materials, the increased supply of sustainable materials has been measurable over the history of LEED. Several MR credits reward use of products that perform well on specific criteria. It is difficult, however, to compare two products that have different sustainable attributes (e.g., cabinets made of wheat husks sourced from all over the country and bound together in resin vs. solid wood cabinets made from local timber). Life-cycle assessment (LCA) provides a more comprehensive picture of materials and products, enabling project teams to make more informed decisions that will have greater overall benefit for the environment, human health, and communities, while encouraging manufacturers to improve their products through innovation.

International Organization for Standardization (ISO) 14040 defines LCA as a "compilation and evaluation of the inputs and outputs and the potential environmental impacts of a product system throughout its life cycle." The entire life cycle of a product or building is examined, the processes and constituents of extraction identified, and their environmental effects assessed; this occurs both upstream, from the point of manufacture or raw materials extraction, and downstream, including transportation, use, maintenance, and end of life. This approach is sometimes called "cradle to grave." Going even further, "cradle to cradle" emphasizes recycling and reuse at the end of life rather than disposal.

Life-cycle approaches to materials assessment began in the 1960s with carbon accounting models. Since then, LCA standards and practices have been developed and refined. In Europe and a few other parts of the world, manufacturers, regulators, specifiers, and consumers in many fields have been using life-cycle information to improve their product selections and product environmental profiles. Until relatively recently, the data and tools that support LCA were lacking in the United States; now, a growing number of manufacturers are ready to document and publicly disclose the environmental profiles of their products, and programs that assist their effort and help users understand the results are available.

LEED aims to accelerate the use of LCA tools and LCA-based decision-making, thereby spurring market transportation and improving the quality of the database. Recognizing the limitations of the life-cycle approach for addressing human health and the ecosystem consequences of raw materials extraction, LEED uses alternative, complementary approaches to LCA in the credits that address those topics.

Required Products and Materials

The scope of the MR credit category includes the building or portion of the building being constructed or renovated. Portions of an existing building that are not part of the construction contract are excluded from MR documentation unless otherwise noted.

Benefits and Issues to Consider

Environmental Issues

By creating convenient recycling opportunities for all building occupants, a significant portion of the solid waste stream can be diverted from landfills. Recycling of paper, metals, glass, cardboards, and plastic reduces the need to extract virgin natural resources. For example, recycling 1 ton of paper prevents the processing of 17 trees and saves 3 cubic yards of landfill space (Oberlin College Recycling Program, 2021). Recycled aluminum requires only 5 percent of the energy required to produce virgin aluminum from bauxite, its raw material form. Diverting waste from landfills can help minimize land, water, and air pollution. An occupant education program that addresses the environmental and financial benefits of recycling can encourage occupants to participate in preserving the environment.

Economic Issues

Many communities sponsor and promote recycling programs to reduce the amount of waste sent to landfills. Community recycling efforts return valuable resources to local production processes and may spur increases in employment in the recycling industry. Community-wide participation results in higher recycling rates and, in turn, more stable markets for recycled materials.

Recycling infrastructure such as storage areas and bins may add to project costs and take up floor area that could be used for other purposes. However, recycling offers significant savings through reduced landfill disposal costs or tipping fees. In larger projects, processing equipment (can crushers, cardboard balers) can minimize the space required for recycling activities. Some recyclables can generate revenue that offsets collection and processing costs.

Related Credits

Project teams seeking an innovation in design credit for educational outreach can create signage and displays to inform building occupants and visitors about on-site recycling.

Core and Shell

Core and Shell project teams should address recycling within tenant guidelines. The tenant guidelines should include information regarding the building's recycling policy and procedures. The project team should encourage activities to reduce and reuse materials before recycling to decrease the volume of recyclables handled.

Implementation

Building owners and designers must determine the best way to create a dedicated recycling collection and storage area that is easily accessible within the building and encourages recycling yet is accessible to the waste hauler. Recyclable material collection and storage space might increase the project footprint in some instances. Consider how recycling activities might affect a building's indoor environmental quality. Activities that create odors, noise, and air contaminants should be isolated or performed during nonoccupant hours. The requirements of this prerequisite do not regulate the size of the recycling area. However, Table 8.1 provides guidelines for the recycling storage area based on overall building square footage, including corridors, elevators, stairwells, and shaft spaces. These guidelines will help the design team determine the appropriate size for recycling facilities to specific building operations.

In places where the materials specified in this prerequisite are not recycled, a building should still have designated space to collect and store those materials in anticipation of recycling infrastructure for the materials becoming available in the future.

Designate and visibly mark central collection and storage areas for recyclables, including paper, cardboard, glass, plastic, and metals. The central collection and storage area should provide easy access for both maintenance staff and collection vehicles; a central collection area designed to consolidate a building's recyclables meets the credit requirements as long as the intent of the credit and the recycling needs of the occupants are met. For projects with larger site areas, it may be possible to create a central collection area that is outside the building footprint or project site boundary. In this case, document how the recyclable materials will be transported to the separate collection area. For projects with landscaping, consider designing an area for collecting plant debris.

Establish recycling collection points within common areas, such as classrooms, break rooms, open offices, and any location where occupants may need to recycle.

Design considerations for a recycling area should include signage to discourage contamination, protection from the elements, and security for high-value materials. Design security for the recyclable collection areas to discourage illegal disposal.

If possible, teach occupants, maintenance personnel, and other building users about recycling procedures. Consider using recycled items, which will reduce the volume of

Commercial Building (sq ft)	Minimum Recycling Area (sq ft)
0 to 5000	82
5001 to 15,000	125
15,001 to 50,000	175
50,001 to 100,000	225
100,001 to 200,000	275
200,001 or greater	500

*This table is referenced from the building design and construction reference guide (Council, U. G. B. (2019)).

TABLE 8.1   Recycling Area Guidelines*

recyclables. For instance, building occupants can reduce the solid waste stream by using reusable bottles, bags, and other containers. Maintenance personal can reduce waste by purchasing cleaning products in bulk or concentrated form. Consider employing cardboard balers, aluminum can crushers, recycling chutes, and other waste management technologies to further improve the recycling program.

Summary

A sustainable building requires policies for responsible construction and materials selection as well as effective waste management. The MR prerequisites and credits establish the foundation for developing, implementing, and documenting these policies. In LEED for schools, many possibilities exist for integrating topics such as resource life cycle, waste reduction, and recycling into the curriculum and for motivating students to become actively involved in conservation efforts at school and in the community.

Operations and building management can effectively reduce a building's overall impact on the environment with waste management programs and purchasing policies that reduce waste and specify less harmful materials and supplies.

Keywords and Definitions

Life-Cycle Costing (also known as Whole-Life Cost): The total cost of ownership over the life of an asset. The concept is also known as life-cycle cost or lifetime cost and is commonly referred to as cradle-to-grave or womb-to-tomb cost.

Life-Cycle Assessment (LCA) (also known as Life-Cycle Analysis): A methodology for assessing environmental impacts associated with all the stages of the life cycle of a commercial product, process, or service.

Vendor of Certified Wood: The company that supplies wood products to contractors or subcontractors for on-site installation. A vendor needs a chain-of-custody (COC) number if it is selling products certified by the Forest Stewardship Council (FCS) that are not individually labeled; this includes most lumber.

Chain of Custody (COC): A tracking procedure for a product from the point of harvest or extraction to its end use, including all successive stages of processing, transformation, manufacturing, and distribution.

Chain-of-Custody Certification: Awarded to companies that produce, sell, promote, or trade forest products after audits verify proper accounting of material flows and proper use of the FSC name and logo. The COC certificate number is listed on invoices for nonlabeled products to document that an entity followed FSC guidelines for product accounting.

Sustainable Forestry: The practice of managing forest resources to meet the long-term forest goal to restore, enhance, and sustain a full range of forest values, including economic, social, and ecological considerations.

Registration Evaluation Authorization and Restriction of Chemicals (REACH): European Union (EU) legislation that requires all chemicals sold in quantity in EU counties to be registered in a central database and prioritized for evaluation and possible avoidance based on their hazard profile. REACH is an alternative compliance path for international projects.

REACH Optimization: When the product contains no ingredients listed on the REACH Authorization or Candidate list, it is valued at least 100 percent of cost for the calculation.

REACH Optimization Detailed Requirement

Option 1: Material ingredient reporting (1 point)

The project has to use at least 20 different permanently installed products from at least 5 different manufacturers that demonstrate the chemical inventory of the product to at least 0.1 percent (1000 pm)

Option 2: Material ingredient optimization (1 point)

The product must document their ingredient optimization for at least 25 percent, based on cost, of the total value of permanently installed products in the project. One of the methodologies recognized by LEED is the REACH optimization. When the product contains no ingredients listed on the REACH Authorization or Candidate list, it is valued at least 100 percent of cost for the calculation.

Option 3: Product manufacturer supply chain optimization (1 point)

Manufacturers have to be engaged in validated and robust safety, health, hazard, and risk programs with independent third party verification of their supply chain.

Additional option:

Products sourced (extracted, manufactured, purchased) within 100 miles of the project site are evaluated at 200 percent of their base contributing cost.

EN 15804: It provides the core rules for the creation of environmental product declarations (EPDs) for building products and materials.

Retrocommissioning: A commissioning process that can be performed on existing buildings to identify and recognize system improvements that make the building more suitable for current use.

Questions

The questions and multiple-choice answers that follow were designed to provide a deeper insight and look into the EQ category.

1. Some of the focuses of the MR category are material conservation, LCA approach, and environmentally preferable materials. What else is on the list of focuses?

   a. Greenhouse gas emission

   b. Waste management

   c. Solid waste disposal

   d. Decreased potable water

2. What are materials that are extracted, manufactured, purchased, consumed, and disposed of with a clear beginning and finite end?

   a. Cradle to gate

   b. Cradle to cradle

   c. Cradle to grave

   d. Grave to cradle

3. Which one of the following is not a strategy to achieve successful waste management of a project?

    a. Waste stream audits

    b. Recycled percentage (15–35) of construction waste from a project with one to two material streams

    c. Compost food waste

    d. Monitor and track waste recycling

4. What is NOT a characteristic of environmentally preferable materials and products?

    a. Known ingredients

    b. Recycled content

    c. Long-lasting, durable, reusable

    d. Shipped materials from oversees with no take-back program

5. Ceiling tiles that can be recycled into a new product at the end of their life would be what type of product?

    a. Cradle to gate

    b. Cradle to grave

    c. Cradle to cradle

    d. Environmental Product Declaration (EPD)

6. Waste management strategies are one of the most important aspects in the MR category. The key to a successful recycling program is to know which materials need to be recycled and to have the appropriate collection and storage areas available. What else is a strategy?

    a. Have a plan to monitor the recycling process

    b. Reuse 90 percent of the building materials

    c. Use local code or jurisdiction

    d. Utilize a Green-e certificate and carbon offsets

7. MR Credit: Construction and Demolition Waste Management awards the project teams 1 to 2 award points for materials recycling of?

    a. 45 percent to 80 percent, respectively

    b. 50 percent to 75 percent, respectively

    c. 50 percent to 50 percent, respectively

    d. No points awarded

8. Waste stream audits in the MR category are a great way to create a baseline for waste diversion improvement. To establish waste diversion goals for a project, the team must identify a minimum of _____ materials targeted for diversion.

    a. Three

    b. Four

    c. Five

    d. Six

9. What is the location valuation factor for a product that is extracted, manufactured, and purchased within 100 miles of the project site?

   a. 100 percent.

   b. 90 percent.

   c. 200 percent.

   d. No valuation factor is designated.

10. How many points are received by reusing historical buildings?

    a. Ten

    b. Six

    c. Two

    d. Three

11. How can an architect make sure that the general contractor understands what products to include in a project?

    a. Conduct weekly meetings

    b. Include the products in the specifications

    c. Hire a LEED consultant

    d. Use a design-bid-build delivery method

12. To help ensure that the products and materials used have preferable life-cycle impacts, the MR category focuses on disclosure and optimization. What does disclosure refer to?

    a. Improving a product or materials to meet specific performance goals

    b. Revealing product information so that optimization becomes more feasible and the process of preferential selection may begin

    c. Optimize existing building resources to reduce the amount of new building materials

    d. Understand the impact of new building materials on overall energy use and the environment through a cradle-to-grave LCA

13. What does a set of specific rules, requirements, and guidelines based on ISO specifications for developing environmental declarations for one or more product categories refer to?

    a. Life-cycle assessments

    b. Environmental product declarations

    c. Health product declarations

    d. Product category rules

14. Which credit rewards the most points among all the other credits in the MR category?

    a. Storage and collection of recyclables

    b. Construction and demolition waste management

    c. Building life-cycle impact reduction

    d. Building Product Disclosure and Optimization (BPDO)-Material ingredients

15. What is wood referred to if it has been procured from well-managed forests?
    a. Sustainable Agriculture Network (SAN)
    b. Forest Stewardship Council (FSC)
    c. REACH optimization
    d. Recycled content

16. During the construction phase, a dispute has been raised between a general contractor and a subcontractor about the quality of materials used in the project. What are the legally binding documents that should describe the details of the materials called?
    a. Specifications
    b. Bidding documents
    c. Legal bidding documents written by a LEED consultant
    d. Arbitration

17. What provides the principles and requirements for type III Environmental Product Declarations (EPDs) for building products?
    a. ISO 14025
    b. ISO 21930
    c. EN 15804
    d. ISO 14040

18. Maintaining at least 50 percent of an abandoned or blighted building receives up to how many points?
    a. Five
    b. Six
    c. Ten
    d. Twenty

19. As a part of the material optimization approaches, what term is applied to products that have fully inventoried chemicals to 100 ppm and that have no benchmark 1 hazards?
    a. Cradle-to-cradle certified
    b. Greenscreen® v1.2 benchmarked
    c. International alternative compliance path-REACH optimized
    d. USGBC-approved program products

20. What influences green design? Select two.
    a. Reducing energy from material transportation
    b. Importing materials from 200 miles without a return program
    c. Performing whole-building LCAs
    d. Performing renovation only on historical buildings

21. A net-zero waste building diverts what percentage of its waste from the landfill?

    a. 99 percent

    b. 95 percent

    c. 100 percent

    d. Depends on the project

22. What term is applied to a product that can be diverted for a new use at the end of its first useful life cycle?

    a. Cradle to grave

    b. Cradle to cradle

    c. Whole-building LCA

    d. a and b

23. How many points does reusing 25 percent to 75 percent of off-site salvaged building materials receive?

    a. Three

    b. Four

    c. Five

    d. Six

24. What is a flow of materials coming from a job site into markets for building materials? It can be specific material categories or a mixture of several materials that are diverted in a specific way.

    a. Life-cycle analysis

    b. Material stream

    c. Composted food waste

    d. Construction and demolition waste planning

25. To qualify for bidding on a project from the general contractor end, the project's owner requires seeing the bidding documents lowering waste-hauling fees and reducing the environmental impact of producing new material. Therefore, what would the LEED consultant recommend having a plan for?

    a. Diverting construction and demolition debris

    b. Purchasing products directly from specific suppliers who have a cradle v3 bronze certificate

    c. Revealing product information so that optimization becomes more feasible and the process of preferential selection may begin

    d. Researching products that are made in factories supporting human health and worker's rights

26. To earn points under Building Product Disclosure and Optimization (BPDP)-Sourcing of Raw materials, project team can use

    a. 25 permanently installed products from at least three manufacturers

    b. 20 permanently installed products from at least three manufacturers

   c. 20 permanently installed products from at least five manufacturers

   d. 25 permanently installed products from at least five manufacturers

27. Construction and demolition waste accounts for about how much of the total solid waste stream in the United States?

   a. 30 percent

   b. 40 percent

   c. 25 percent

   d. 50 percent

28. The MR Prerequisite: Storage and Collection of Recyclable requires project teams to offer recycling options for which two of the following items?

   a. Batteries, electronic waste

   b. Glass, plastic, and metals

   c. Mixed paper and corrugated cardboard

   d. Mercury-containing lamps

29. An owner is deciding which type of insulation to select. Besides performance and adaptability, what else should the owner consider?

   a. Leverage points

   b. Embodied energy

   c. Reuse of old insulation

   d. Salvage materials

30. A LEED professional is looking at selecting interior materials that have lifetime positive effect. Select two from the below materials or items that have less environmental consequence.

   a. Reused furniture

   b. Flooring shipped from over 300 miles

   c. Demountable interior walls

   d. Products that have an EPD

31. An owner is building an apartment complex with 1000 units. What recommendations can the owner get to save on material costs and conserve materials?

   a. Use prefabricated frames

   b. Use paints with low volatile organic compounds (VOCs)

   c. Buy materials from a local store

   d. Follow the Material Safety Data Sheet (MSDS) safety manual to avoid safety hazards

32. Door materials that can be recycled into new products at the end of their useful life would be considered what type?

   a. Cradle to gate

   b. Active

c. Cradle to cradle

d. Cradle to grave

33. What is a greenhouse gas that is produced from landfills that is 28 times more potent than carbon dioxide?

a. Land Fill Gas (LFG)

b. Methane

c. Neon

d. Argon

34. A project team discusses the selection of products that meet sustainable design objectives. Which of the following meets these criteria?

a. Process continues through postoccupancy

b. Process finished when construction is complete

c. Process limited in constrained optimization

d. Linear process

35. The EPA recommends solid waste management by ranking approaches. Which of the following is the least preferred approach?

a. Recycling

b. Waste to energy

c. Source reduction

d. Reuse

36. Select two of the following that help by completing an LCA.

a. Reducing the amount of Request For Information (RFI)

b. Reducing the waste of materials used

c. Using change order reduction

d. Selecting the best materials to fit the project

37. What are the most-effective ways of increasing material conservation? Select two.

a. Historic building reuse

b. Renovation of abandoned or blighted buildings

c. Health Product Declarations (HPDs)

d. Product Category Rule (PCR)

38. What is the purpose of manufacturers disclosing materials' ingredients?

a. To determine if the products are locally produced

b. To choose more reasonable products

c. To make better decisions on product selections

d. To help select team members

39. What is the stuff that ends up on the factory floor and is repurposed into something new rather than trashed called?

    a. Postconsumer recycled materials

    b. Preconsumer recycled materials

    c. Long-lasting materials

    d. Product-specific materials

40. By performing a whole-building LCA, a new construction building can conduct an LCA of the project's structure to demonstrate a minimum of 10 percent reduction. This process can be compared with a reference building in at least _____ of six impact measures.

    a. Two

    b. Three

    c. Four

    d. Five

41. The greenscreen v1.2 benchmark is used in which of the following credit categories?

    a. Sustainable Site

    b. Location and Transportation

    c. Materials and Resources

    d. Energy and Atmosphere

42. Which of the following helps identify products, so manufacturers can verify, extract, and source materials in responsible manner.

    a. Health Product Declaration (HPD)

    b. Environmental Product Declaration (EPD)

    c. Corporate Sustainability Report (CSR)

    d. ISO 26000

43. If a product is found in a landfill and used for the same purpose or another purpose, what is it called?

    a. Comingled material

    b. Reused material

    c. Recycled content material

    d. Cradle-to-grave material

44. All except one of the following are third-party certification types. Which one is not a certification type?

    a. Forest Stewardship Council

    b. Energy Star

    c. WaterSense

    d. Waste stream audit

45. Building products are selected based on what in regard to the environmental impacts? Select all that apply.

    a. Product size impact

    b. Impact of these products on society

    c. Environmental impact

    d. Cost impact

46. What is the economic benefit of reusing furniture?

    a. Reduces demand for virgin materials

    b. Provides more design options to the design team

    c. Reduces the tax fee and lowers the purchasing price

    d. Provides rustic and classy looks

47. Part of the administrative side of a project is preparing documents for keeping waste out of landfills by reducing, reusing, and recycling materials and products. What does that represent?

    a. Owner's project requirement

    b. Construction waste management plan

    c. Commissioning plan

    d. Storage and collection recycling plan

48. A LEED professional was consulted to provide recommendations for design configuration and flexibility. Which two options did the LEED professional recommend?

    a. Modular and movable casework

    b. Demountable partitions

    c. Solar panels

    d. Interior open space garden

49. To maintain an indoor facility that is more user friendly and to eliminate any potential hazard, the facility manger should enforce use of which of the following?

    a. Material Safety Data Sheet

    b. MERV 8 filters

    c. Health product disclosure

    d. Building life-cycle impact reduction

50. Which is a good example of adaptive reuse?

    a. Historic building renovation

    b. Renovation of an old prison facility into a museum

    c. Demolition and building of a new office space

    d. None of the above

51. Material health reutilization, renewable energy and carbon management, water stewardship, and social fairness are all reward achievements offered by which of the following?

    a. Cradle-to-cradle certificate

    b. Life-cycle assessment

    c. Life-cycle costing

    d. Greenscreen

52. A LEED professional is developing a plan for construction debris that is nonrecyclable waste. Which is the best practice?

    a. Recycling

    b. Using waste to energy

    c. Comingling or separation

    d. Performing optimization

53. The raw material report can be a self-declared report by the manufacturer or a report by which of the following?

    a. Third-party-verified CSR

    b. Life-Cycle Assessment (LCA) or Life-Cycle Costing (LCC) report

    c. Location valuation factor report

    d. REACH criteria report

54. How is a baseline created for waste diversion improvement?

    a. Conduct waste stream audits

    b. Survey the building occupants

    c. Count the number of recycling bins

    d. Develop a sustainable materials policy

55. An owner wants to build an eco-friendly facility that demands fewer raw materials. Which project might that be?

    a. A multistory building built on a greenfield

    b. A dense missed-use neighborhood

    c. A medical facility in suburb

    d. Community gardens

56. What is life-cycle assessment used for?

    a. To determine indoor and outdoor water consumption

    b. To reduce the VOC level

    c. To help assess life-cycle costing

    d. To understand the trade-offs of material selection and energy performance

57. LEED projects use a document that transparently communicates the environmental performance or impact of any product or material over its lifetime. This document indicates information about global warming and greenhouse gas emissions. What is it called?

a. Greenhouse gas emission declaration

b. EPD

c. Carbon offset document

d. Low-emitting materials document

58. A team is trying to decide on door selections through the door contents and what they are made from. How should the team do that?

a. Use MSDSs

b. Use weight and size

c. Check the manufacturer's reputation

d. Use an LCA

References

Oberlin College recycling program. (2021). Recycling Facts. http://www.oberlin.edu/recycle/facts.html. Accessed March 2020.

U.S. Environmental Protection Agency (EPA). (April 2022). Municipal Solid Waste: Fact and Figures. Accessed September 2022. https://www.epa.gov/landfills/municipal-solid-waste-landfills#whatis.

Wasmi, H. A., & Castro-Lacouture, D. (2016). Potential impacts of BIM-based cost estimating in conceptual building design: a university building renovation case study. In *Construction Research Congress 2016* (pp. 408-417).

DGXI European Commission: Final Report (February 1999), Construction and Demolition Waste Management Practices, and Their Economic Impacts. Report by Symonds in association with ARGUS COWI and PRC Bouwcentrum. Accessed September 2022. http://www.resol.com.br/textos/Construction%20and%20demolition%20Waste%20management%20part%201.pdf.

EPA, Resource Conservation and Recovery (RCR, 2012): A guide to developing and implementing greenhouse gas reduction programs. Local government climate and energy strategy guides. Accessed September 2022. https://19january2017snapshot.epa.gov/sites/production/files/2015-08/documents/resource_conservation_and_recovery_a_guide_to_developing_and_implementing_greenhouse_gas_reduction_programs.pdf.

UGBC, green building fact: LEED V4, reference guide for interior design and construction. Accessed September 2022. https://www.usgbc.org/guide/idc.

City of Seattle LEED Projects Analysis, Seattle LEED Projects 2000-2011: Executive Summary. Accessed September 2022. https://www.seattle.gov/documents/Departments/OSE/COS_LEED_project_analysis2000-11.pdf.

ISO 14040 international standard environmental management, life cycle assessment. principles and framework (Geneva, Switzerland: international organization for standardization 2006. Section 5: Methodological framework, article 5.3: life cycle inventory analysis (LCI). Accessed September 2022. https://www.iso.org/obp/ui/#iso:std:iso:14040:ed-2:v1:en.

Council, U. G. B. (2019). *LEED reference guide for building design and construction.* US Green Building Council.

Answers

1. **b** The Materials Resources category focuses on a multiattribute approach and simultaneously examines a broad range of environmental and human health factors across the project's entire life cycle.

2. **c** Option c is the definition of cradle to grave.

3. **b**

4. **d** Option d does not support an eco-friendly endeavor. The cost of transportation and embodied energy are big factors that damage the environment.

5. **c** Read more on cradle to cradle.

6. **a** Monitoring the cycling process is important and part of the waste management strategies.

7. **b**

8. **c**

9. **c**

10. **b**

11. **b**

12. **b**

13. **d**

14. **c** The scorecard indicates all the reward points, so option c has more points than the others.

15. **b** The FSC provides a third-party certification.

16. **a**

17. **b** This question is about the third-party certification that provides the framework for LCAs and EPDs.

18. **b**

19. **b**

20. **a & c**

21. **c**

22. **b**

23. **c**

24. **b** The materials stream is explained as seen in the question statement.

25. **a**   The LEED professional recommends using a waste management plan, which is diversion, and demolition debris is included.

26. **c**

27. **b**   Here, 40% is the percentage that is recommended.

28. **a & d**

29. **b**   Embodied energy is the total energy required for the extraction, processing, manufacture, and delivery of building materials to the building site. A leverage point in a system is where a small change can lead to a large change in results.

30. **a & c**   Option b, selecting materials from a great distance is not supported by the MR category. For option d, EPD is used to compare products and select those that are less harmful. A product may perform better in some environmental categories than others as listed on the EPD.

31. **a & c**   When selecting building materials, they are three factors to be considered to facilitate selection:

    1. Performance: How will the materials perform compared with others? For insulation, what are the best insulating properties?

    2. Adaptability: Can the insulation be reused in the future when the building is repurposed?

    3. Embodied energy: This is the amount of energy used during the life cycle of the materials, starting from harvesting, processing, manufacturing, storing, shipping, and installing.

32. **c**   A closed system is the best system; cradle to cradle is an example of it. Cradle to cradle describes materials that are recycled into a new product at the end of its life.

33. **b**   Solid waste disposal in landfills causes a release of methane, which is more harmful than CO_2.

34. **a**   Option a is part of the integrative process, while all the other options fall under the conventional process.

35. **b**   The EPA recommends four approaches professionals in sustainability could use to reduce waste. These approaches are ranked as follows: source reduction, reuse, recycling, and waste to energy.

36. **b & d**   On the one hand, reducing RFIs and change orders has nothing to do with LCA. On the other hand, LCA helps select the best materials and components that fit the project for the long term and reduces waste.

37. **a & b**   Options c and d are product attributes and not methods of material conservation.

38. **c**   Option c is the best option because the material's ingredients help team members select the best product (i.e., help the team make more informed decisions). Other options are not related to a material's ingredients.

39. **b**   Check out the definition of preconsumer recycled materials.

40. **b**   At least three of the six impact measures must be met. One of the three impact measures must be global warming potential.

41. **c**

42. **c** Option c is the definition of a CSR.

43. **b**

44. **d** Options a, b, and c are all third-party certifications required to develop a sustainable materials policy. Option d is not a certification.

45. **b, c, d** The question asks about the triple bottom line and its criteria. The triple bottom line does not consider product size as a deciding factor, which is option a. The triple bottom line is based on three elements: society (people), environment, and profit (cost).

46. **c** The question asks about the economic benefit, which is the c option. Option a is for environmental benefit, so is b. Option d is for aesthetic purposes.

47. **b** A commissioning plan is a document that outlines the organization and resource allocation requirements. A storage and collection recycling plan is a required item, and it is for providing dedicated recyclable material collection and storage areas for occupants. Option b, which is a construction waste management plan, contains an administrative plan for waste management.

48. **a & b** For configurable and flexible design, a and b options are the best choices. The other options are for different design purposes.

49. **a** MERV 8 is a type of filter for an HVAC system. MSDSs provide workers procedures to safely handle a product during fabrication, installation, or a life safety event. The health product disclosure is an impartial tool for the accurate reporting of product contents and the relationship of each ingredient to human and ecological health.

50. **b** Adaptive reuse is the process of reusing an existing building for a purpose other than the one for which it was originally built or designed. Option a, historic renovation, is just a renovation. Demolition is an adaptive reuse approach.

51. **a** Cradle-to-cradle certified is a product certification program for assessing and constantly improving products by requiring product ingredients to be disclosed to an independent, accredited "cradle to cradle" assessor.

52. **b** Option a, recycling, does not work because the problem statement indicates that the materials are nonrecyclable. Option c does not help because the materials are nonrecyclable, so either comingling or separating cannot help earning credit. Option d is not applicable. Option b, which is waste to energy, is the best option because the nonrecyclable materials can turned into energy.

53. **a** Do not confuse EPDs or LCA with raw material sourcing. A CSR helps identify products/manufacturers that have been verified to be extracted or sourced in a responsible manner.

54. **a** A waste stream audit is a process that takes a sample of all waste produced by a facility. It provides a clear indication of the work the organization needs to accomplish in recycling and waste diversion.

55. **b** The greater the density of a project, the more environmentally friendly it is. More dense buildings require less infrastructure.

56. **d**   Life-cycle assessment is defined by the USGBC as an evaluation of the environmental effects of a product from cradle to grave.

57. **b**   EPD includes information about the environmental impact, such as ozone depletion, water pollution, or greenhouse gas emission.

58. **d**   The purpose of a life-cycle assessment is to compare various materials and identify those with significantly negative impact over the building and environment.

CHAPTER 9

Indoor Environmental Quality (EQ)

The Indoor Environmental Quality (EQ) category rewards decisions made by project teams about Indoor Air Quality (IAQ) and Thermal, Visual, and Acoustic Comfort. Green buildings with good indoor environmental quality protect the health and comfort of building occupants. According to the U.S. Environmental Protection Agency (EPA, 2001), high-quality indoor environments also enhance productivity, decrease absenteeism, improve the building's value, and reduce liability for building designers and owners. This category addresses the myriad design strategies and environmental factors—air quality, lighting quality, acoustic design, control over one's surroundings—that influence the way people learn, work, and live.

The relationship between the indoor environment and the health and comfort of building occupants is complex and still not fully understood. Local customs and expectations, occupant's activities, and the building's site, design, and construction are just a few of the variables that make it difficult to quantify and measure the direct effect of a building on its occupants. Therefore, the EQ category balances the need for prescriptive measures with more performance-oriented credit requirements. For example, source control is addressed first, in a prerequisite, and a later credit then specifies an IAQ assessment to measure the actual outcome of those strategies.

The EQ chapter combines traditional approaches such as ventilation and thermal control with merging and monitoring for user-determined contaminants (enhanced IAQ strategies credit), requirements for lighting quality (interior lighting credit), and advanced lighting metrics (daylight credit).

Intent of Indoor Air Quality Category

The intent of the IAQ category is to contribute to the comfort and well-being of building occupants by establishing minimum standards for IAQ.

Requirements of the Indoor Air Quality Category

The requirements for EQ must be met for both Ventilation and Monitoring. Ventilation and Monitoring are the main two sections, which consist of mechanically ventilated spaces and naturally ventilated spaces.

Ventilation

1. Mechanically Ventilated Spaces

   a. **Option 1: American Society of Heating, Refrigeration and Air-Conditioning Engineers (ASHRAE) Standard 62.1-2010**

   For mechanically ventilated spaces (and for mixed-mode systems when the mechanical ventilation is activated), determine the minimum outdoor air intake flow for mechanical ventilation systems using the ventilation rate procedure from ASHRAE 62.1-2010 or a local equivalent that is more stringent.

   Meet the minimum requirements of ASHRAE Standard 62.1-2010, Section 4-7, Ventilation for Acceptable Indoor Air Quality (with errata), or a local equivalent, whichever is more stringent.

   b. **Option 2: CEN Standards EN 15251-2007 and CEN Standards EN 13779-2007**

   Projects outside the United States may instead meet the minimum outdoor air requirements of Annex B of the European Committee for Standardization (Comité Européen de Normalisation, CEN) Standard EN 15251-2007, indoor environmental input parameters for design and assessment of energy performance of building addressing IAQ, thermal environment, lighting, and acoustics and meet the requirements of CEN Standard EN 13779-2007, Ventilation for Nonresidential Buildings, Performance Requirements for Ventilation and Room Conditioning Systems, Excluding Section 7.3, Thermal Environment; 7.6 Acoustic Environment; A.16 and A.17.

2. **Naturally Ventilated Spaces**

   For naturally ventilated spaces (and for mixed-mode systems when the mechanical ventilation is inactivated), determine the minimum outdoor air opening and space configuration requirements using the natural ventilation procedure from ASHRAE Standard 62.1-2021 or a local equivalent, whichever is more stringent. Confirm that natural ventilation is an effective strategy for the project by following the flow diagram in the Chartered Institution of Building Service Engineers (CIBSE) application manual AM10, March 2005, Natural Ventilation in Nondomestic Buildings, and meet the requirements of ASHRAE Standard 62.1-2010, Section 4, or a local equivalent, whichever is more stringent.

3. **All Spaces**

   The IAQ procedure defined by ASHRAE Standard 62.1-2021 may be used to comply with this prerequisite.

Monitoring

1. **Mechanically Ventilated Spaces**

   For mechanically ventilated spaces (and for mixed-mode systems when the mechanical ventilation is activated), monitor outdoor air intake flow as follows:

   - For variable air volume systems, provide a direct outdoor measurement device capable of measuring the minimum outdoor air intake flow. This device must measure the minimum outdoor air intake flow with an accuracy of ±10 percent of the design minimum outdoor airflow rate, as defined by the ventilation requirements above. An alarm must indicate when the outdoor airflow value varies by 15 percent or more from the outdoor airflow set point.

- For constant-volume systems, balance outdoor airflow to the design minimum outdoor airflow rate defined by ASHRAE Standard 62.1-2021 (with errata), or higher, install a current transducer on the supply fan, an airflow switch, or similar monitoring device.

2. Naturally Ventilated Spaces

For naturally ventilated spaces (and for mixed-mode systems when the mechanical ventilation is inactivated), comply with at least one of the following strategies:

- Provide a direct exhaust airflow measurement device capable of measuring the exhaust air. This device must measure the exhaust airflow with an accuracy of ±10 percent of the design minimum exhaust airflow rate. An alarm must indicate when the airflow values vary by 15 percent or more from the exhaust airflow set point.

- Provide automatic indication devices on all natural ventilation openings intended to meet the minimum opening requirements. An alarm must indicate when any one of the openings is closed during occupied hours.

- Monitor carbon dioxide (CO_2) concentrations within each thermal zone. CO_2 monitors must be between 3 and 6 feet (900 and 1800 millimeters) above the floor and within the thermal zone. CO_2 monitors must have an audible or visual indicator or alert the building automation system if the sensed CO_2 concentration exceeds the set point by more than 10 percent. Calculate appropriate CO_2 set points using the methods in ASHRAE 62.1-2010.

According to the EPA (2001), Americans spend an average of 90 percent of their time indoors, so the quality of the indoor environment has a significant influence on their well-being, productivity, and quality of life. The EPA reported that pollutant levels of indoor environments may run two to five times—and occasionally more than 100 times—higher than outdoor levels. Similarly, the World Health Organization reported in its second edition of *Air Quality Guidelines for Europe* edition that most of an individual's exposure to air pollutants comes through inhalation of indoor air. Following EPA releases of the "Reducing Risk" (1990) and "Indoor Air in Large Buildings" (2002) that designated indoor air pollution as a top environmental risk to public health, assessing and managing indoor pollutants have become the focus of integrated governmental and private efforts. Recent increases in building-related illnesses and "sick building syndrome," as well as an increasing number of related legal cases, have further heightened awareness of IAQ among building owners and occupants. Strategies to improve indoor environmental quality have the potential to reduce liability for building owners, increase the resale value of the building, and improve the health of building occupants.

For schools and schoolchildren, indoor environmental quality issues are even more urgent. From promoting best practice led by the American Academy of Allergy, Asthma and Immunology (AAAAI, 2004), it is known that many pollutants cause an adverse health reaction in the estimated 7 million children and adolescents who suffer from asthma, contributing to 14.7 million days of absence in schools each year. In fact, based on the work of the Asthma and Allergy Foundation of America (n.d.), asthma is the leading chronic illness and chief cause of absenteeism among school-age children. In the United States, more than 56 million children and 7.1 million teachers spend a considerable amount of time in school buildings according to the U.S. Census Bureau (Davis, 2008). The indoor environmental quality in these buildings can have a significant effect on the health and well-being of students and staff, as well as on the quality and effectiveness of the learning environment.

Compared with adults, children are at greater risk of exposure to and possible illness from environmental hazards because of their greater sensitivity during development and growth. In addition to upper respiratory infections and asthma, continuous exposure to pollutants can cause symptoms such as nausea, dizziness, headaches, lethargy, inattentiveness, and irritation of the eyes, nose, and throat. Continuous exposure to hazardous substances can also lead to learning disabilities, cancers, and illnesses caused by damage to the nervous system.

In addition to health and liability concerns, productivity gains also drive indoor environmental quality improvements. With employees' salaries a significant cost in any commercial building, it makes good business sense to keep staff healthy and productive by improving and maintaining the quality of the indoor environment. The potential annual savings and productivity gains from improved indoor environmental quality in the United States are estimated at $6 billion to $14 billion from reduced respiratory disease, $1 billion to $4 billion from reduced allergies and asthma, $10 billion to $30 billion from reduced sick building syndrome symptoms, and $20 billion to $160 billion from direct improvements in worker performance that are unrelated to health (OSHA, 2011).

Over the past 20 years, research and experience have improved our understanding of what is involved in attaining high indoor environmental quality and revealed manufacturing and construction practices that can prevent many indoor environmental quality problems. The use of better products and practices has reduced the potential liability for design team members and building owners, increased market value for buildings with exemplary indoor environmental quality, and boosted the productivity of building occupants. In a case study included in the 1994 publication, "Greening the Building and the Bottom Line," the Rocky Mountain Institute highlighted how improved indoor environmental quality increased worker productivity by 16 percent netting a rapid payback on the capital investment.

This credit category addresses environmental concerns relating to indoor environmental quality; occupants' health, safety, and comfort; energy consumption, air change effectiveness, and air contaminant management. Discussed next are important strategies for addressing these concerns and improving indoor environmental quality.

Improving Ventilation

Actions that affect employee attendance and productivity will affect an organization's bottom line. One study estimated a 283 percent return on investment associated with increased ventilation in less than 6 months (Damiano and Dougan, 2003).

Specifying building systems will provide a high level of IAQ. Increased ventilation in buildings may require additional energy use, but the need for additional energy can be mitigated by using heat-recovery ventilation or economizing strategies. IAQ design can help take advantage of regional climate characteristics and reduce energy costs. In regions with significant heating or cooling loads, for example, using exhaust air to heat or cool the incoming air can significantly reduce energy use and operating costs.

Managing Air Contaminants

Protecting the indoor environment from contaminants is essential for maintaining a healthy space for building occupants. Several indoor air contaminants should be reduced to optimize tenants' comforts and health. There are three basic contaminants:

1. **Environmental tobacco smoke** (or secondhand smoke) is both the smoke given off by ignited tobacco products and the smoke exhaled by smokers. Environmental tobacco smoke contains thousands of chemicals, more than 50 of which are carcinogenic, as stated by the U.S. Department of Health and Human Services, National Institute of Health, and National Cancer Institute. Exposure to environmental tobacco smoke is linked to an increased risk of lung cancer and heart diseases in nonsmoking adults and is associated with increased risk of sudden infant death syndrome and asthma, bronchitis, and pneumonia in children. Smoking should be eliminated in all indoor building spaces and limited to designated outdoor areas.

2. **Carbon dioxide** concentrations can be measured to determine and maintain adequate outdoor air ventilation rates in buildings. CO_2 concentrations are an indicator of air change effectiveness. Elevated levels suggest inadequate ventilation and possible buildup of indoor air pollutants. CO_2 levels should be measured to validate indications that ventilation rates need to be adjusted. Although relatively high concentrations of CO_2 alone are not known to cause serious health problems, they can lead to drowsiness and lethargy in building occupants (Prill, 2000).

3. **Particulate matter** in the air degrades the indoor environment. Airborne particles in indoor environments include lint, dirt, carpet fibers, dust, dust mites, mold, bacteria, pollen, and animal dander. These particles can exacerbate respiratory problems such as allergies, asthma, emphysema, and chronic lung disease (Goren et al., 1999). Air filtration reduces the exposure of building occupants to these airborne contaminants, and high-efficiency filters greatly improve IAQ. Protecting air handling systems during construction and flushing the building before occupancy further reduce the potential for problems to arise once the building is occupied.

Specifying Less Harmful Materials

Preventing indoor environmental quality problems is generally much more effective and less expensive than identifying and solving them after they occur. A practical way to prevent indoor environmental quality problems is to specify materials that release fewer and less harmful chemical compounds. Adhesives, paints, carpets, composite wood products, and furniture with a low level of potentially irritating off-gassing can reduce occupants' exposure and harm. Appropriate scheduling of deliveries and sequencing of construction activities can reduce material exposure to moisture and absorption of off-gassed contaminants.

Allowing Occupants to Control Desired Settings

Working with building occupants to assess their needs will help improve building efficiencies. Providing individual lighting controls and area thermostats can improve occupants' comfort and productivity and save energy. Individual controls enable occupants to set light levels appropriate to tasks, time of day, personal preferences, and individual variations in visual acuity. Individual thermostats enable them to more accurately meet their heating and cooling needs during different seasons.

Providing Daylight and Views

Daylight reduces the need for electric lighting, which lowers energy use and thereby decreases the environmental effects of energy production and consumption. Natural daylight also increases occupants' productivity and reduces absenteeism and illness. Studies have shown that providing daylight and exterior views can measurably increase academic performance in schools. Courtyards, atria, clerestory windows, skylights, interior light shelves, exterior fins, louvers, and adjustable blinds, used alone or in combination, are effective strategies to achieve deep daylight penetration. The desired amount of daylight depends on the tasks in a given space. Daylight buildings often have several daylight zones with differing target light levels. In addition to light levels, daylighting strategies affect interior color schemes, direct beam penetration, and integration with the electric lighting system.

Building occupants with access to outside views have an increased sense of well-being, leading to higher productivity and increased job satisfaction. Important considerations for providing views include building orientation, window size and spacing, glass section, and locations of interior walls.

Core and Shell Adoption

In their design and construction, core and shell projects can affect IAQ in two ways. First, the design and construction teams can influence the quality of interior spaces, such as lobbies, central circulation areas, and building cores. Second, core and shell design and construction decisions can directly affect the indoor environmental quality of tenant spaces outside the control of the Core and Shell category submittal. Examples include ventilation design and careful design consideration for tenants' ability to optimize daylight and views. Design and construction teams in Core and Shell projects should consider how their decisions could enable tenant fit outs to deliver high indoor environmental quality to building occupants.

Schools

Reduce Background Noise and Provide Good Acoustics in Schools

Creating a high-performance acoustic environment is important in learning spaces because human communication is a primary foundation of learning. Minimizing background noise and optimizing acoustics through careful design and material choices enable effective teacher-to-student and student-to-student communication.

Compliant Space Types for Indoor Environmental Quality Credits

The following list of three types of spaces identifies school spaces considered to be regularly occupied for applicability to indoor environmental quality credits. In these spaces, daylight, views, thermal comfort, or acoustics affect the quality of occupants' regular use. Leadership in Energy and Environmental Design (LEED) will evaluate expectations to these classifications on a case-by-case basis for spaces with atypical uses or those in which the strategies required for compliance may compromise the function of the space.

1. Regularly Occupied Space: Classroom and Core Learning

The category of regularly used space consists of spaces that are used for at least 1 hour per day for educational activities where the primary functions are teaching and learning. Table 9.1 shows these spaces.

2. Other Regularly Occupied Spaces

The category of other regularly occupied spaces includes all nonlearning spaces that are used by occupants for 1 hour or more per day to perform work-related activities. Table 9.2 shows these spaces.

3. Space Not Regularly Occupied

Spaces considered not regularly occupied are those that occupants pass through and those that are not regularly used for at least 1 hour per day. Table 9.3 shows these spaces.

Art	Gymnasium	Physical education
Band	Instructional technology	Physical lab
Biology lab	Instrument instruction	Vocational arts
Chemistry lab	Language lab or art	Voice instruction
Chorus	Library	Classroom
Media center	Computer lab	

TABLE 9.1 Regularly Occupied Spaces Primarily for Teaching and Learning

Administrative conference room	Counselor's office	School nurse's treatment room
Administrative office	Faculty office	School security office
Administrative staff room	Faculty workroom	Staff dining room
Cafeteria, cafetorium	Maintenance staff room	Counseling conference room
Natatorium	School nurse's office	

TABLE 9.2 Other Regularly Occupied Spaces Primarily for Nonlearning Spaces

Administration waiting	Locker room	Students' activities room
Auditorium	Main entrance	Students' locker area
Backstage area	Receiving area	Corridor
Stage	Greenhouse	Stairs

TABLE 9.3 Spaces Not Regularly Occupied

Summary

Ensuring excellent indoor environmental quality requires the joint efforts of the building owner, design team, contractors, subcontractors, and suppliers. To provide optimal indoor environmental quality, automatic sensors and individual controls can be integrated with the building systems to adjust temperature, humidity, and ventilation. Sensors can measure building CO_2 levels and indicate the need for increased outdoor airflow to eliminate a high level of volatile organic compounds (VOCs) and other air contaminants. Other indoor environmental quality issues addressed by the LEED New Construction, Core and Shell, and Schools rating system include daylighting and lighting quality, thermal comfort, acoustics, and access to views. These issues all have the potential to enhance the indoor environment and optimize interior spaces for building occupants.

Keywords and Definitions

Building Envelope: The exterior surface of a building's construction: the walls, windows, roof, and floor.

ASHRAE Standard 55-2010: It specifies conditions for acceptable thermal environments and is intended for use in design, operation, and commissioning of buildings and other occupied spaces.

There are six variables of thermal comfort according to ASHARE standard 55-2010. The standard 55 is oriented toward providing thermal comfort, addressing the following six variables: Metabolic Rate, Clothing Insulation, Air Temperature, Radiant Temperature, Air Speed, and Humidity.

Sheet Metal and Air Conditioning Contractors' National Association (SMACNA): It is an international trade association with more than 4500 contributing contractor members in 103 chapters throughout the United States, Canada, Australia, and Brazil. SMACNA standards and manuals address all facets of the sheet metal and HVAC industry, including duct construction and installation, indoor air quality, energy recovery, roofing and architectural sheet metal, welding, and commissioning.

Vision Glazing: It is the quality view of building occupants' line of sight to outdoor (looking from inside to outside) or indoor (looking from outside to inside). Example, a building occupant sitting inside a building and looking outside without object(s) obstruction. Line of sight is important for building occupants to connect with the outside natural environment. Another example, a building user may request more privacy from visitors looking through to the inside by partially blocking the line of sight.

Quality view in the IEQ category can be achieved by providing vision glazing for 75 percent of all regularly occupied floor area (Council, U.G.B (2013)). The vision glazing must not be obstructed by frits, fibers, patterned glazing, or type of distortion. In addition to giving building users natural outdoor connection, 75 percent of all regularly occupied floor area must at least have two of the following:

1. Provide multiple lines of sight to vision glazing in different directions at least 90 degrees apart.
2. Provide views that include at least two of 1) flora, fauna, or sky; 2) movement; and 3) objects at least 25 feet from the exterior glazing.
3. Provide unobstructed views located within the distance of three times the head height of the vision glazing.

Daylighting: The practice of placing windows, skylights, other openings, and reflective surfaces so that sunlight can provide effective internal lighting. Particular attention is given to daylighting while designing a building when the aim is to maximize visual comfort or to reduce energy use.

Infiltration: Air leakage into conditioned spaces through cracks and interstices in ceiling, floors, and walls from unconditioned spaces or the outdoors (ASHRAE 62.1-2010).

Thermal Zone: A space or collection of spaces having similar space-conditioning requirements and the same heating and cooling set point and is the basic thermal unit (or zone) used in modeling the building. A thermal zone will include one or more spaces.

U Value: The measure of heat flow through materials that separate the building façade, slab, or roof from the exterior environment in units.

Vapor Barrier: Any material used to prevent moisture penetration through wall, celling, and floor assemblies, and potential condensation that can result from differences between a building's interior and exterior temperatures.

Volatile Organic Compounds: Organic chemicals that have a high vapor pressure at room temperature. High vapor pressure correlates with a low boiling point, which relates to the number of the sample's molecules in the surrounding air, a trait known as volatility.

Questions

The questions and multiple-choice answers that follow were designed to provide a deeper insight and look into the EQ category.

1. What does ASHRAE 55-2010 define?
   a. Ventilation system design
   b. Range of indoor thermal environmental conditions acceptable to a majority of occupants
   c. The specific conditions for acceptable thermal environments and the use of design, operation, and commissioning
   d. Area of ground that the building occupies as defined by its perimeter
   e. Measurement used to determine airflow rates

2. Which is a manual that defines the acceptable process for natural ventilation of nondomestic buildings?
   a. CIBSE AM10
   b. American National Standards Institute (ANSI)
   c. Closed system
   d. Credit Interpretation Request (CIR)

3. What must the increased outdoor air ventilation rates to all occupied spaces above the minimum rates as determined in EQ Prerequisites: Minimum IAQ Performance be?
   a. 15 percent
   b. 30 percent

    c. 25 percent

    d. 5 percent

4. According to the EPA, headache, eye, nose, throat irritation, dry cough, dizziness, and nausea are all symptoms of what?

    a. Lack of access to the daylight

    b. Demand response (DR)

    c. Influenza

    d. Sick building syndrome

5. What are naturally ventilated spaces measured by?

    a. CIBSE AM10

    b. Minimum efficiency reporting value (MERV) 13

    c. IAQ management plan

    d. SMACNA

6. EQ Prerequisite Site-Environmental Tobacco Smoke Control requires smoking to be allowed outside the building in designated smoking areas. Where should these areas be located?

    a. More than 25 feet from all entries

    b. At least 25 feet from the main door of the building

    c. At least 25 feet from all entries and outdoor air intakes

    d. Within 25 feet of all windows and air exhaust

7. To improve interior lighting per EQ Credit: Interior Lighting, a project team must meet the requirements by providing individual lighting controls with at least three lighting levels for at least _____ of the individual occupant's spaces?

    a. 100 percent

    b. 90 percent

    c. 50 percent

    d. 10 percent

8. To achieve EQ Credit: Quality Views, a project team must achieve a direct line of sight to the outdoor via vision glazing for _____ of all regularly occupied floor areas.

    a. 90 percent

    b. 75 percent

    c. None

    d. 10 percent

9. Which three strategies improve the IAQ?

    a. Improving air ventilation, which reduces health issues and increases building performance

    b. Preventing sick building syndrome

c. Setting minimum IAQ performance

d. Developing Type III environmental declaration programs

10. How many points are available in the EQ category?

a. 13

b. 11

c. 16

d. 33

11. What standard establishes minimum indoor quality?

a. ASHRAE 55-2010

b. ASHRAE 62.1-2010

c. ASHRAE 99-2010

d. MERV

12. Suppose that a working space is evaluated based on its occupancy. A mechanical room is unoccupied most of the month (say 29 days of the month it is unoccupied). On the last day of the month (30th day), the space is utilized by the mechanical crew for the entire day (8 hours) for maintenance purposes. What is considered the occupation designation of the space?

a. Nonregularly occupied

b. Regularly occupied

c. Unoccupied

d. Partially occupied

13. What is the largest reason for IAQ problems?

a. Cooling and heating

b. Inadequate ventilation

c. Low-emitting materials

d. Lack of waste stream audits

14. How many prerequisites does the EQ category have according to the score card?

a. Three

b. Two

c. One

d. None

15. To prevent signs of sick building syndrome, the U.S. building codes and professional standards generally require the ventilation per person in an office space to be which value?

a. 15–20 Cubic feet per Minute (cfm)

b. 20–30 Cubic feet per Minute (cfm)

c. 15–20 cubic in inch square (c/in²)

d. 13 MERV (Minimum Efficiency Reporting Values)

16. In an arid climate, a project team wants to allow building occupants to have temperature control all the time. What should the team do?

    a. Provide adjustable-height partitions

    b. Provide thermostats

    c. Provide adjustable air diffusers

    d. Install more windows

17. What can the circadian rhythms be identified as?

    a. The establishment of higher quality indoor air in the building after construction and during occupancy

    b. A process used to remove airborne contaminants from a building caused by the off-gassing of furniture, paints, adhesives, sealants, carpets, and other building materials by flushing out all the interior air several times with outdoor air

    c. A daily cycle of biological activity generated by an internal clock that is based on a 24-hour period and influenced by regular variations in the environment, such as the alternation of night and day

    d. A rating system that ranges from 1 to 16 for a filter made for a mechanical building ventilation system that filters particulates and airborne contaminants out of the air

18. Low-emitting materials for the interior and exterior of buildings are organized into seven categories for VOCs, each with a different threshold of compliance. Select three.

    a. Flooring

    b. Cabinets and casework

    c. Roof

    d. Composite wood

    e. Furniture

19. A data center main floor and storage areas are considered as?

    a. Occupied space

    b. Unoccupied space

    c. Sometimes occupied

    d. Empty spaces

20. To meet the requirements of EQ Credit: Interior Lighting, what percentage of individual lighting control must be provided?

    a. Two lighting levels for at least 90 percent of individual occupants' spaces

    b. Three lighting levels for at least 95 percent of individual occupants' spaces

    c. Lighting levels for at least 90 percent of individual occupants' spaces

    d. Three lighting levels for at least 90 percent of individual occupants' spaces

21. In order to properly capture dirt and particulates, a permanent entryway system should be installed and be at least _____ feet long in the primary direction of travel.

    a. 30

    b. 10

    c. 50

    d. 5

22. Sealing off the construction area, protecting ductwork, and ventilating the construction areas are all measures of protecting building occupants from inhaling dust, chemicals, and the particles in the air due to construction. This procedure is performed under which guideline?

    a. SMACNA

    b. MERV 13

    c. Stack drive ventilation approach

    d. ASHRAE 62.1-2010

23. What is the controlled admission of natural light into a space to reduce or eliminate electric lighting called?

    a. Daylighting

    b. Glazing

    c. Task lighting

    d. SRI (Solar Reflectance Index)

24. The CIBSE provides a standard manual to measure which of the following?

    a. Mechanical ventilation

    b. Natural ventilation

    c. Daylight and interior lighting

    d. Thermal comfort

25. The project team wants to promote minimizing pest problems and human and environmental exposure to pesticides. What does the LEED consultant advise using?

    a. Circadian rhythms approach

    b. Adaptive and native plants

    c. An organized program of nonchemical strategies

    d. Chemical in significant amount

26. In arid climate, a project team is challenged to provide ventilation control. What should the team do?

    a. Provide natural ventilation

    b. Provide natural ventilation with operable windows

    c. Provide adjustable air diffusers

    d. Provide movable furniture

27. Why is establishing a green cleaning policy important?

    a. To encourage more participation of the building's facility manager

    b. To make the janitorial staff more effective

    c. To reduce levels of chemical, biological, and particulate contaminants

    d. To reduce the costs of cleaning a building

28. The EQ Credit: Green Cleaning: Custodial Effectiveness Assessment; EQ Credit: Green Cleaning: Products and Materials; and EQ Credit: Green Cleaning: Equipment are all credits that can grant you points. What are these credits of?

    a. LEED BD+C (Building Design and Construction) rating system

    b. LEED O+M (Maintenance and Operation) rating system

    c. LEED ID+C (Interior Design and Construction)

    d. LEED ND (Neighborhood and Development)

29. Building orientation is one of the vital approaches to control excessive light. Which option provides occupants more control over excessive light?

    a. Shading devices

    b. High wall participations

    c. Task lights in multiple areas

    d. Reconfigurable building layout

30. To implement a good IAQ strategy, which optional credits are recommended? Select three.

    a. Use low-emitting materials

    b. Use disposable janitorial products

    c. Conduct an annual audit accordance with the Association of Physical Plant Administrators

    d. Conduct environmental tobacco smoke control

31. According to the National Institute of Occupational Safety and Health, the reason for most indoor air problems is inadequate ventilation. What is the percentage of inside contamination?

    a. 10 percent

    b. 15 percent

    c. 5 percent

    d. 4 percent

32. A project team is working on reducing glare inside a building. What strategy is most effective and less costly?

    a. Using tasked lighting

    b. Providing low-wall partitions

    c. Installing interior curtains

    d. Installing double-pane windows

33. During operation and maintenance, why is it recommended to conduct occupant surveys in LEED projects?

    a. It is part of minimum project requirements.

    b. It is required by the Owner Project Requirement.

    c. A survey provides vital information on underperforming areas.

    d. It is required by ASHRAE 55-2010.

34. To bring a sense of vitality and connection to the outdoors that improves health, production, and happiness during the design phase, what should the project team do?

    a. Locate private offices on the perimeters of the building

    b. Centrally locate private offices and open offices on the perimeter of the floor plan

    c. Use large curtain walls and keep the private offices on the perimeters

    d. Use native and adapted plants on the building's surrounding

35. Select three primary factors of thermal comfort.

    a. Humidity

    b. LED (light-emitting diode) lighting

    c. Air movement

    d. Reflective panels

    e. Air temperature

36. A project team is working on IAQ, so what should the team pay attention to?

    a. Interior lighting

    b. Daylighting during the day

    c. Quality view

    d. CO_2 level

37. An owner undecided on a decision for selecting a new flooring with a low VOC versus a more reasonable high VOC. Which of the following would be impacted by the decision?

    a. Materials disclosure and optimization

    b. Construction and demolition waste management planning

    c. Storage and collection of recyclables

    d. Indoor air quality

38. Part of the LEED project's requirements is to prohibit smoking near _____. Select two.

    a. Air intakes

    b. Outside seating 25 feet from the building

    c. Operable windows

    d. Trees and newly planted flowers

39. What does occupant control over lighting and thermal comfort help?

    a. Improve mechanical ventilation

    b. Improve IAQ strategies

    c. Improve low-emitting materials

    d. Improve performance and productivity

40. The higher the ventilation rate, the cleaner the indoor air will be as all of the toxins are swept out and replaced by fresh outdoor air. What determines a good ventilation level in a space for good IAQ?

    a. Type of the ventilation system, such as a mechanical system

    b. Type of the ventilation system, such as mechanical and natural systems

    c. The time of day the space is used

    d. The number of occupants and type of activities

41. Name one trade-off for an open layout concept.

    a. Low air quality and poor air movement

    b. Low ventilation quality

    c. Acoustic problems

    d. Poor thermal comfort

42. Entryway systems are one of the IAQ strategies that is used to promote occupant comfort, well-being, and productivity. Select the type of materials you would use for an entryway system method.

    a. Carpet shipped from overseas with a return back program

    b. Rubber mat that captures dirt and particulates entering the building

    c. Hardwood flooring from a supplier within a 100-mile diameter

    d. Recycled carpet

43. Who is responsible for conducting occupant surveys and making corrections to building performance if there is a need?

    a. Architect

    b. Commissioning authority

    c. Engineer

    d. Owner

44. A building with good performance includes low VOC materials, at least a 10-foot long entryway system, MERV 13 and above, and air intakes located near an operating machine. Which of these approaches is not suitable for a good building environment?

    a. Entryway system

    b. Low VOC materials

    c. MERV 13

    d. Air intake

45. A building is being constructed in an area where the air quality is not the best. What should the project team do to provide good IAQ? Select two.

    a. Use more natural ventilation

    b. Install operable windows and a timer to shut windows when needed

    c. Install permanent entryway systems

    d. Allow an outside smoking area at least 25 feet from all entries

46. Increasing ventilation and providing more access to an outside view would affect the triple bottom line in terms of which of the following?

    a. Decrease occurrence of periodic costs

    b. Decrease global warming

    c. Increase occupant performance

    d. None of the above

47. What must a building owner require to maintain building performance at a high level? Select two.

    a. Open and close windows when needed

    b. Hire a good custodial staff

    c. Monitor sensor performance and provide regular calibration

    d. Routinely replace air filters

48. What measures must a contractor do during construction to protect building occupants later from inhaling dust, chemicals, and other particles in the air due to construction? Select three.

    a. Sealing off the construction area

    b. Protecting ductwork

    c. Ventilating the construction areas

    d. Implementing integrated pest management

49. Buildings relying partially or fully on natural ventilation allow their occupants to have control over the amount of air entering the building. What are some of these factors that allow occupants to have that kind of control?

    a. Sensors to control daylight and interior lights

    b. Sensors to control thermal comfort

    c. Operable windows with manual or automatic shutoff

    d. Movable furniture

50. An owner wants to certify his building to the highest LEED certification possible. The owner designated a small space for smoking inside the building. How can the owner certify the building?

    a. By installing proper signage indicating the designated hours during the days that occupants can smoke

    b. By installing several windows to the room so they can provide enough ventilation

    c. By knowing the building does not meet the environmental tobacco smoke control prerequisite

    d. By installing proper signage indicating the designated hours during the days that occupants can smoke and also limiting the smokers to a specific number

51. What products need to be checked for low-emitting VOCs? Select two.

    a. Chairs and tables

    b. Wall paint and sealants

    c. Outside seating benches

    d. Building sidings

52. There is a method of pest management that protects human health and the surrounding environment and improves economic returns through the most effective, least risk option. What is this the definition of?

    a. Occupant comfort survey

    b. Greenguard certification

    c. Integrated pest management

    d. Sick building syndromes

53. Once the design phase is passed by the project participants and approved by the owner, changes are typically costly on the budget. During construction, what can the owner change without affecting the budget significantly?

    a. Building orientation

    b. Building share and layout

    c. Recycled furniture with low-emitting contents

    d. Window ratio to the floor

54. Including natural ventilation in the design of the building impacts which credits? Select two.

    a. Enhanced IAQ strategies

    b. Green power

    c. Light pollution reduction

    d. Optimal energy performance

55. What does a team need to consider during the design phase when thinking about a floor plan?

    a. Incorporating recycled content materials as much as possible

    b. Sound transfer and noise

    c. Designated smoking areas

    d. Minimum efficiency reporting value

56. Including task lighting and increasing ventilation levels in a new building have what effect on the triple bottom line?

    a. Increased occupant performance

    b. Decreased occupant performance

    c. Creation of an unhealthy environment

    d. Increased practice expense

57. Why would a project team use low-emitting materials?

    a. Provide a more affordable solution

    b. Protect the health of construction personnel and building occupants

    c. Provide easy-to-use and install type of materials

    d. Provide low-emitting materials that should be used on LEED projects

References

American Academy of Allergy, Asthma, and immunology (AAAAI). (2004). Pediatric Asthma: Promoting Best Practice: Guide for Managing Asthma in Children. American Academy of Pediatrics. Accessed November 2020.

American Society of Heating, Refrigeration and Air-Conditioning Engineers. ASHRAE 62.1-2010. ASHRAE Standards: Ventilation for Acceptable Indoor Air Quality. http://arco-hvac.ir/wp-content/uploads/2016/04/ASHRAE-62_1-2010.pdf.

Asthma and Allergy Foundation of America. Asthma Facts and Figures. https://www.aafa.org/asthma-facts/. Accessed November 2020.

Centers for Disease Control and Prevention, National Center for Health Statistics. Asthma Prevalence, Health Care Use and Mortality, 2002. https://www.cdc.gov/nchs/data/hestat/asthma/asthma.htm#:~:text=In%202002%2C%204%2C261%20people%20died,adults%20aged%2018%20and%20over. Accessed September 2022.

Damiano, L., and D. Dougan. (2003). The Big Carrots: Productivity and Health. Ebtron.

Occupational Safety and Health Administration (OSHA): Indoor Air Quality in Commercial and Institutional Buildings. Occupational Safety and Health Administration, U.S. Department of Labor. OSHA 3430-04, 2011. Accessed September 2022. https://www.osha.gov/sites/default/files/publications/3430indoor-air-quality-sm.pdf.

Goren, A., Hellmann, S., Gabbay, Y., & Brenner, S. (1999). Respiratory problems associated with exposure to airborne particles in the community. *Archives of Environmental Health: An International Journal, 54*(3), 165-171. Accessed September 2022. https://www.tandfonline.com/doi/abs/10.1080/00039899909602255.

Prill, R. (2000). Why measure carbon dioxide inside buildings. *Published by Washington State University Extension Energy Program WSUEEP07, 3.* Accessed September 2022. https://www.energy.wsu.edu/documents/co2inbuildings.pdf.

Romm, J. J., & Browning, W. D. (1994). Greening the building and the bottom line. *Rocky Mountain Institute. Snowmass, Colorado.* Accessed January 2022. https://www.terrapinbrightgreen.com/wp-content/uploads/2015/05/Greening_the_Building_and_the_Bottom_Line.pdf.

Davis, J. W., & Bauman, K. J. (2008). School Enrollment in the United States: 2006. Population Characteristics. Current Population Reports. *US Census Bureau.* Accessed March 2021. https://eric.ed.gov/?id=ED520727.

United States. Environmental Protection Agency. Office of Policy, & Evaluation. Office of Policy Analysis. (1987). *Unfinished business: A comparative assessment of environmental problems* (Vol. 2). US Environmental Protection Agency, Office of Policy Analysis, Office of Policy Planning and Evaluation. Accessed March 2021. https://books.google.com/books?hl=en&lr=&id=sA3izkPk1tMC&oi=fnd&pg=PP11&dq=U.S.+Environmental+Protection+Agency+(EPA).+(1987).+Unfinished+Business:+A+Comparative+Assessment+of+Environmental+Problems.+Washington,+DC:+U.S.,+EPA,.+1987+&ots=q6i4tA1TUj&sig=0ZojUzHgJja2wAf7OCu8WXpr-yE#v=onepage&q=U.S.%20Environmental%20Protection%20Agency%20(EPA).%20(1987).%20Unfinished%20Business%3A%20A%20Comparative%20Assessment%20of%20Environmental%20Problems.%20Washington%2C%20DC%3A%20U.S.%2C%20EPA%2C.%201987&f=false.

U.S. Environmental Protection Agency (EPA). (1990). Reducing Risk: Setting Priorities and Strategies for Environmental Protection. Washington, DC: U.S. EPA, 1990.

U.S. Environmental Protection Agency (EPA). (2001). Healthy Buildings, Healthy People: A Vision for the 21st Century. Accessed May 2020. https://www.epa.gov/indoor-air-quality-iaq/healthy-buildings-healthy-people-vision-21st-century.

U.S. Environmental Protection Agency (EPA). (2002). Indoor Air in Large Buildings. Accessed February 2019. https://www.ashrae.org/technical-resources/bookstore/indoor-air-quality-guide.

ASHRAE Standard 62.1-2010: Ventilation for Acceptable Indoor Air Quality. ANSI/ASHRAE Standard 62.1-2010. Accessed November 2021. http://arco-hvac.ir/wp-content/uploads/2016/04/ASHRAE-62_1-2010.pdf.

CEN Standards EN 15251-2007: Indoor environmental parameters for design and assessment of energy performance of buildings addressing indoor air quality, thermal environment, lighting, and acoustics. Accessed September 2022. https://standards.iteh.ai/catalog/standards/cen/92485123-bf64-40e3-9387-9724a642eae8/en-15251-2007.

CEN Standards EN 13779-2007: Ventilation for non-residential buildings-performance requirements for ventilation and room-conditioning systems. Europeans Standard. ICS 91.140.30. Accessed September 2022. http://www.freedom2choose.org.uk/wp-content/uploads/2017/06/EC_Standard_For_Ventilation.pdf.

Chartered Institution of Building Service Engineers (CIBSE) (2005): natural ventilation in non-domestic buildings. CIBSE applications manual AM10. Carbon Trust. Accessed June 2019. http://arco-hvac.ir/wp-content/uploads/2018/04/Natural_ventilation_in_non_domestic.pdf.

World Health Organization (WHO), Regional Office for Europe Copenhagen: Air Quality Guidelines for Europe, second edition. WHO Regional Publications, European Series, No 91 (2000). Accessed October 2019. https://www.euro.who.int/data/assets/pdf_file/0005/74732/E71922.pdf.

Council, U. G. B. (2013). *LEED reference guide for building design and construction*. US Green Building Council. Accessed May 2021.

Answers

1. **c** Choosing c because it is the definition of American Society of Heating, Refrigeration and Air-Conditioning Engineers (ASHRAE) Standard 55-2010. Option d is area measure, and option e has to deal with cubic feet per minute (cfm), which is the unit measure of airflow. Option a is for ventilation design governed by ASHRAE 62.1-2010 and option b is irrelevant.

2. **a** The Chartered Institution of Building Services Engineers (CIBSE) Application Manual AM 10 is the manual standard that must be followed for design of natural ventilation. The other options are irrelevant.

3. **b** To fulfill the requirements of EQ Prerequisite: Minimum IAQ Performance Ventilation Rate, the increase of the outdoor zone must be at least 30%.

4. **d**

5. **a** CIBSE AM10 is a manual standard and contains a flow diagram, which helps to design natural ventilation.

6. **c**

7. **b**

8. **b**

9. **a, b, c** Option d belongs to the MR (Materials and Resources) category.

10. **c** Check out the scorecard.

11. **b** ASHRAE 62.1-2010 is the standard for ventilation that is acceptable for indoor air quality (IAQ). ASHRAE 55-2010 is for thermal comfort, ASHRAE 99-2010 does not exist, and MERV (minimum efficiency reporting value) is the filtering rating.

12. **b** Even though the space is not utilized almost the entire month, one full day use is considered regularly occupied. Keep in mind that if a space is occupied for less than an hour, then this space is always nonregularly occupied.

13. **b** Cooling and heating are less problematic than ventilation because occupants can always control them through thermostats or by cracking windows open. Low-emitting materials could be a problem due to volatile organic compounds (VOCs); however, ventilation can help flush out VOC effects. Lack of waste stream audits has to deal with material waste outdoors.

14. **b** Check out the scorecard.

15. **a** The project team must comply with the code and professional standards set forth by government authorities.

16. **b** Because the arid climate is hot and dry, outside air is not very useful for occupants' comfort. Providing thermostats is more ideal to ensure proper ventilation and IAQ. Adjustable air diffusers are less effective than thermostats. Option a is irrelevant, and option d does not contribute to the "temperature control."

17. **c**

18. **a, d, e** I recommend memorizing the seven categories for low-emitting materials of VOC products.

19. **b** Remember that if people spend less than an hour in a space, the space is considered nonregularly occupied. The data center main floor is mainly for equipment, so are storage areas.

20. **d** The question is meant to trick you and push you to pay more attention to the required percentage and its relevancy to the individual spaces.

21. **b**

22. **a**   Sheet Metal and Air Conditioning Contractors' National Association (SMACNA) provides the IAQ guidelines for control measures for occupied buildings under construction.

23. **a**   Using daylight definitely reduces the need for artificial lighting. More glazing helps to bring in more daylight, but the question here aims for a more specific answer. Task lighting is not related to natural lighting, and SRI is Solar Reflection Index.

24. **b**

25. **c**   Integrated Pest Management (IPM) is a credit program in the Leadership in Energy and Environmental Design (LEED) Maintenance and Operation (O+M) rating system for the building and grounds within the project boundary. It encourages elimination of the use of chemical products.

26. **c**   Option c is more relevant than all the other options. This question is tricky and meant to measure your understanding and help you practice choice elimination.

27. **c**   Green cleaning policy is an approach to reduce chemical-based materials inside the building, and it is not about cost. Some chemical materials are more expensive than green cleaning products.

28. **b**

29. **a**   Shading devices control excessive outside light entering the building. Options b and d are helpful, but only partially and are not as effective as shading devices. Option c is irrelevant.

30. **a, b, c**   All three options are applicable; option d is a prerequisite for outdoor smoking.

31. **b**

32. **c**   Double-pane windows help in thermal control not glare. Interior curtains are more common and less expensive than other choices.

33. **c**   This approach is very effective and more common in LEED projects.

34. **b**   Option b promotes synergy between daylighting and quality view. To achieve this type of synergy, storage areas, private offices, and conference rooms should be located in the center of the building. This allows more daylight and a quality view for the rest of the open floor.

35. **a, c, e**   There are six thermal (according to SHRAE 55-20210): clothing, air movement, metabolic rate, humidity, air temperature, surface temperature.

36. **d**   All options are reasonable, but CO_2 has the most effect on the IAQ.

37. **d**   Even though high VOC is more cost reasonable then low VOC, the IAQ will be negatively affected, so sometimes we have to think about the trade-offs.

38. **a & c**

39. **d**   This question is meant to practice choice elimination and see which option fits best.

40. **d**   For a good ventilation system, CO_2 is a big concern that should be addressed. When the number of occupants is high, then the level of CO_2 is high. The type of activities are also important, so people just sitting or sleeping is significantly different from people working out.

41. **c** In a work environment, people may talk over each other, which creates distraction. Also, playing music while others do want to listen it is another problem.

42. **b**

43. **d** Surveys typically are conducted after occupancy, so the owner is a reasonable option. In large facilities, the facility manager is responsible.

44. **d** An air intake should be located away from any source of pollution.

45. **c & d** Since the outside air is not good quality, then the least the team can do is implement options c and d.

46. **c** The effect is positive here, so think about the question more before making a selection.

47. **c & d** The c and d options are more effective and efficient than the other two options.

48. **a, b, c** The key words here are "during construction," not after occupancy, which refers to the IPM method.

49. **c** Only option c is related to natural ventilation. All other options are unused or irrelevant to natural ventilation.

50. **c** Allowing smoking inside does not satisfy the prerequisite Environmental Tobacco Smoke Control. If a building does not meet a prerequisite, then even if it meets all the credits, the building will not receive the certification.

51. **a & b** Low-emitting materials contain seven product categories, options a and b included. The rest of the options are not included in the seven categories.

52. **c**

53. **c** Option c is the best approach to not make any design changes. All others will affect not only the budget, but also the schedule and many more.

54. **a & d** Natural ventilation saves on energy use and sweeps out any indoor air contamination.

55. **b** Sound transfer or acoustics in general is a common problem in the open floor concept, so a project team must consider the trade-off.

56. **a**

57. **a**

Innovation (IN) and Regional Priority (RP)

Innovation (IN)

Sustainable design strategies and measures are constantly evolving and improving. New technologies are continually introduced to the marketplace, and up-to-date scientific research influences building design strategies. The purpose of the Innovation (IN) Leadership in Energy and Environmental Design (LEED) category is to recognize projects for innovative building features and sustainable building practices and strategies.

Occasionally, a strategy results in building performance that greatly exceeds what is required in an existing LEED credit. Other strategies may not be addressed by any LEED prerequisites or credit but warrant consideration for their sustainability benefits. In addition, LEED is most effectively implemented as part of a cohesive team, and this category addresses the role of a LEED-accredited professional in facilitating that process.

Intent of Innovation

The intent of the IN category is to encourage projects to achieve exceptional or innovative performance.

Requirements of Innovation

Project teams can use any combination of innovation, pilot, and exemplary performance strategies.

Option 1: Innovation (1 point)

Achieve significant, measurable environmental performance using a strategy not addressed in the LEED green building rating system

Identify the following:

- Intent of the proposed IN credit
- Proposed requirements of compliance
- Proposed submittals to demonstrate compliance
- Design approach or strategies used to meet the requirements

and/or

Option 2: Pilot (1 point)
Achieve one pilot credit from the U.S. Green Building Council's (USGBC's) LEED Pilot Credit library

and/or

Option 3: Additional strategies

Innovation (1–3 points)

- Defined in Option 1

Pilot (1–3 points)

- Meet the requirements of Option 2

Exemplary performance (1–2 points)

- Achieve exemplary performance in an existing LEED prerequisite or credit that allows exemplary performance, as specified in the LEED reference guide. An exemplary performance point is typically earned for achieving double the credit requirements of the next incremental percentage threshold.

Sustainable design strategies and measures are constantly evolving and improving. New technologies are continually introduced to the marketplace, and up-to-date scientific research influences building design strategies. The purpose of this LEED category is to recognize projects for innovative building features and sustainable building practices and strategies. Occasionally, a strategy results in building performance that greatly exceeds what is required in an existing LEED credit. Other strategies may not be addressed by any LEED prerequisite or credit but warrant consideration for their sustainability benefits. In addition, LEED is most effectively implemented as part of an integrated design process, and this category addresses the role of a LEED-accredited professional in facilitating that process.

Implementing New Technologies and Methods

As the building design and construction industry introduces new strategies for sustainable development, opportunities leading to additional environmental benefits will continue to emerge. Opportunities that are not currently addressed by LEED for New Construction, Schools, or Core and Shell may include environmental solutions specific to a particular location, condition, or region. With all sustainable strategies and measures, it is important to consider related environmental impacts. Project teams must be prepared to demonstrate the environmental benefits of innovative strategies and are encouraged to pursue opportunities that provide benefits of particular significance. Project teams can earn exemplary performance points for implementing strategies that result in performance that greatly exceeds the level of scope required by an existing LEED prerequisite or credit.

Benefits and Issues to Consider

Sustainable design comes from innovative strategies and thinking. Institutional measures to reward such thinking like the achievement of any credits benefit our environment. Recognition of the exceptional will spur further innovation. Every LEED for New Construction, Core and Shell, and School holds ideas for innovation in design and strategies.

Implementation

Exemplary Performance Strategy

Exemplary performance strategies result in performance that greatly exceeds the level of scope required by existing LEED New Construction, Core and Shell, and Schools.

Innovative Strategies

Innovative strategies are those that are not addressed by any existing LEED credits. Only those strategies that demonstrate a comprehensive approach and have significant, measurable environmental benefits are applicable.

Timeline and Team

Innovation ideally begins at a project's conception, but it can enter at any step of the process and come from any member of the project team. Open-mindedness, creativity, and rigor in follow-through are the critical ingredients. Options for innovation may come from the spheres of the technological (e.g., an inventive wall section for climate control) or the general (e.g., educational outreach measures). Thus, team members with a variety of skills and interests will be able to contribute to the achievement of this credit.

Regional Priority (RP)

Because some environmental issues are particular to a locale, volunteers from USGBC chapters and the LEED International Roundtable have identified distinct environmental properties within their areas and the credits that address those issues. These regional priority credits encourage project teams to focus on their local environmental priorities.

The USGBC established a process that identified six RP credits for every location and every rating system within chapter or country boundaries. Participants were asked to determine which environmental issues were most salient in their chapter area or country. The issues could be naturally occurring (e.g., water shortages) or human-made (e.g., polluted watersheds) and could reflect environmental concerns (e.g., water shortages) or environmental assets (e.g., abundant sunlight). The areas, or ozone, were defined by a combination of priority issues (e.g., an urban area with an impaired prioritized credits addresses the important issues of given locations). Because each LEED project type (e.g., a data center) may be associated with different environmental impacts, each rating system has its own RP credits.

The ultimate goal of RP credits is to enhance the ability of LEED project teams to address critical environmental issues across the country and around the world.

Intent of Regional Priority

The intent of RP credits is to provide an incentive for the achievement of credits that address geographically specific environmental, social equality, and public health priorities.

Requirements for Regional Priority

Up to four of the six regional priority credits can be earned. These credits have been identified by the USGBC regional council and chapters as having additional regional importance for the project's region. A database of RP credits and their geographic applicability is available on the USGBC website (www.usgbc.org/rpc). One point is awarded for each RP credit achieved, up to a maximum of four.

The main difference in RP credits between LEED 2009 and LEED V4.1 is that credit zones are no longer identified by U.S. ZIP Codes, but through a Geographic Information Systems (GIS)-based program. This created RP credit zones that are more environmentally specific and not solely based on physical location. LEED Online allows project teams to enter their physical coordinates, or if they don't know them, to easily find the coordinates by entering their address. When registering a project, LEED Online uses an intuitive interface that allows teams to easily find their project's location. Then, all eligible RP credits will automatically populate.

Questions

The questions and multiple-choice answers that follow were designed to provide a deeper insight and look into the Innovation (IN) and Regional Priority (RP) categories.

1. Which category encourages achieving exceptional performance?
    a. Regional Priority
    b. Innovation
    c. Integrative Process
    d. Location and Transpiration
2. Exemplary performance and pilot credits are available in which category?
    a. Regional Priority
    b. Innovation
    c. Integrative Process
    d. Energy and Atmosphere
3. What is the purpose of the Pilot Credit library?
    a. To test new or revised LEED credits
    b. To provide an incentive for achievement of credits that address geographically specific environmental, social, equity, and public health priorities
    c. To reflect the part of the project surrounding and public outreach concerning regional design
4. Innovation credits
    a. Are only available to Gold certified projects
    b. Are only available to Certified certified projects
    c. Are in every rating system
    d. Have been used since 2014
5. A project team pursuing an exemplary performance for an IN credit must
    a. Achieve all prerequisites and credits in a single category
    b. Achieve all prerequisites and double credits in a single category

    c. Achieve double the credit requirements or the net incremental parentage threshold of an existing credit

    d. Achieve all of the credits in the five categories: EQ (Environmental Quality), EA (Energy and Atmosphere), MR (Materials and Resources), WE (Water Efficiency), SS (Sustainable Site)

6. The IN category is meant for which choice? Select two.

    a. Project teams exceeding the thresholds listed in the existing credits and achieving points for exemplary performance

    b. Encouraging the use of products and materials for which life-cycle information is available and that have environmentally, economically, and socially preferable life-cycle impacts

    c. Pursuing new credit ideas and/or exceeding existing credits

    d. Reducing the peak amount of energy used in the operation and maintenance of a building and the energy use of the occupants

7. The Innovation Credit: Innovations points (1–5) can be achieved through which three of the following choices?

    a. Regional Priority

    b. Innovation

    c. Pilot credit

    d. Exemplary performance

    e. Demand response

8. Which strategies and ideas project teams can use in projects to earn innovation credits? Select three.

    a. Outdoor recreation areas and playground equipment for children

    b. Eliminating all the outdoor seating areas

    c. On-site community exercise facilities

    d. Signs posted at hallways to encourage use of the stairs

    e. Signs posted at staircases to encourage using elevators

9. The number of Innovation Credit: Innovation credit points is

    a. Six

    b. Five

    c. One

    d. Two

10. The purpose of the LEED Pilot Credit library is to

    a. Test new or revised LEED credits

    b. Launch a preselected plan by the project team

    c. Achieve geographically specific environmental credits

    d. Use trial and error

11. The intent of Pilot Credit: Medical and Process Equipment Efficiency is
    a. To encourage walkability
    b. To maximize opportunities for integrated, cost-effective adoption of green design and construction strategies
    c. To reduce energy consumption by using efficient medical and other equipment
    d. To use the most updated medical equipment

12. An exemplary performance credit is given when
    a. Ensuring ample daylight in the room
    b. Achieving an increase in ventilation (in EQ Credit: Enhanced Indoor Air Quality Strategies) of 60 percent or more above the baseline (which is 30 percent)
    c. Achieving a permanent location on existing land
    d. Using an addenda database

13. How is an RP determined?
    a. Using a program based on GIS (geographic information system)
    b. Driving around and checking out landmarks
    c. Making assumptions on each region separately and then verifying
    d. Counting RP credits automatically

14. An RP credit is
    a. Innovation
    b. LEED-Accredited Professional
    c. Specific Credit
    d. Exemplary performance

15. Regional priority credits are specific to which choice?
    a. The city where the project is
    b. LEED categories
    c. LEED project type
    d. LEED rating system

16. What is a LEED rating system credit that is designed to test new and revised LEED credit language, alternative compliance paths, and new or innovative green building technologies and concepts called?
    a. Exemplary performance credit
    b. Innovative credit
    c. Pilot credit
    d. Regional priority credit

17. Smart growth is incredibly supported in the built environment. All except which choice is an example of smart growth?
    a. Building a community center facility for the surrounding community far from existing developments
    b. Building a multifamily homes on a previously developed campus

c. Building a retail location and office building on a former gas station

d. Building in a neighborhood development community

18. An owner is deciding on choices for developing a new office building. Which site is preferred?

a. In a high-density area

b. On a greenfield area

c. In an area with three retail stores

d. Near a water body

Answers

1. **b**

2. **b**

3. **a**

4. **c**

5. **c**   Option c is the correct option; check out the definition of exemplary performance.

6. **a**

7. **b, c, d**

8. **a, c, d**   Options b and e are not items that encourage green practice.

9. **b**

10. **a**

11. **c**   Check out Leadership in Energy and Environmental Design (LEED) certification standards and references for the correct answer.

12. **b**   Option b is the correct answer. Check out the exemplary performance for the Environmental Quality (EQ) category.

13. **a**   The correct option is a. A program based on GIS (geographic information system) allows for environmental issues to be mapped and creates credit zones and is not simply based on physical location.

14. **c**   The correct option is c because each location has its own particular climate, population density, and regulation. Thus, Regional Priority is a specific credit.

15. **c**   The same answer is given in Question 17.

16. **c**   Review the definition of pilot credit.

17. **a**   Option a does not support smart growth because the goal is to build on previously developed lands and near more developments and infrastructure to avoid disturbing undisturbed lands.

18. **a**   Options b and d are not acceptable. Option c is a good option, but LEED requires more diversity of use.

LEED Terminologies and Definitions

Leadership in Energy and Environmental Design (LEED) program is a third-party green building certification program and an international symbol of excellence in the design, construction, and operation of high-performance green buildings and neighborhoods. It encourages and accelerates adoption of sustainable building and community development practices through the creation and implementation of a green building benchmark that is voluntary, consensus-based, and market-driven. Comprehensive and flexible, LEED is applicable to buildings at any stage in their life cycles. New construction, the ongoing operations and maintenance of an existing building, and a significant tenant retrofit to a commercial building are all addressed by LEED rating systems. The rating systems and their companion reference guides help teams make the right green building decisions for their projects through an integrated process, ensuring that building systems work together effectively. Through a consensus-based process, the rating systems are continually evaluated and regularly updated to respond to new technologies and policies and to changes in the built environment. In this way, as yesterday's innovation becomes today's standard of practice, United States Green Building Council (USGBC) and LEED continue to push forward market transformation.

A project must adhere to the LEED's Minimum Program Requirements (MPRs), or possess minimum characteristics in order to be eligible for certifications. These requirements define the categories of building that the LEED rating systems were designed to evaluate, and taken together serve three goals: (1) give clear guidance to customers, (2) protect the integrity of the LEED program, (3) reduce challenges that occur during the LEED certification process. The MPRs will evolve over time in tandem with the LEED rating systems. In order to be eligible for certification in any LEED rating system, projects must comply with each associated MPR. The Green Building Certification Institute (GBCI) reserves the right to revoke LEED certification from any LEED project upon gaining knowledge of non-compliance with any applicable MPRs. If such a circumstance occurs, no registration or certification fees paid to GBCI will be refunded.

149

LEED Neighborhood Development (ND) Only

A site that is qualified as a LEED neighborhood development site must meet any of the following four conditions. Once one condition is met, then the site can be granted the ND certification.

a. At least 75 percent of its boundary borders parcels that individually are at least 50 percent previously developed, and that in aggregate are at least 75 percent previously developed.

b. The site, in combination with bordering parcels, forms an aggregate parcel whose boundary is 75 percent bounded by parcels that individually are at least 50 percent previously developed, and that in aggregate are at least 75 percent previously developed.

c. At least 75 percent of the land area, exclusive of rights-of-way, within ½ mile (800 meters) of the project boundary is previously developed.

d. The lands within ½ mile (800 meters) of the project boundary have a pre-project connectivity of at least 140 intersections per square mile (54 intersections per square kilometer).

The circulation network itself does not constitute previously developed land; it is the status of property on the other side of the segment of circulation network that matters. For conditions (a) and (b) above, any fraction of the perimeter that borders a water body is excluded from the calculation.

The terminologies below are applicable to the majority of LEED projects across disciplines. Readers are encouraged to review and study them to be familiar with the technical vocabularies and standard definitions. Additionally, several of the technical terminologies are explained with elaboration of their requirements and applicability to project sites. All the terminologies are sorted in alphabetical order for easy reference.

A

Abandoned Property   Property left behind intentionally and permanently when it appears that the former owner does not intend to come back, pick it up, or use it. One may have abandoned the property of contract rights by not doing what is required by the contract. However, an easement and other land rights are not abandoned property just because of nonuse. Abandoned land is defined as land not being used at the present time but that may have utilities and infrastructure in place.

Adapted Plant   Vegetation that is not native to a particular region but has characteristics that allow it to live in the area. Adapted plants do not pose the same problems as invasive species.

Added Antimicrobial Treatment   A substance added to a product (e.g., paint, flooring) to kill or inhibit the growth of microorganisms. Some products, such as linoleum, exhibit natural antimicrobial properties. Despite current practice, science has not proven that antimicrobial treatments reduce infection transfer in building finishes more effectively than standard cleaning procedures. Also known as added microbial agent. See the U.S. EPA fact sheet, "Consumer Products Treated with Pesticides" (https://www.epa.gov/safepestcontrol/consumer-products-treated-pesticides).

Adjacent Site   A site having at least a continuous 25 percent of its boundary bordering parcels that are previously developed sites. Only consider bordering parcels, not intervening rights-of-way. Any fraction of the boundary that borders a water body is excluded from the calculation.

Alternative Daily Cover (ADC)   Material other than earthen material placed on the surface of the active face of a municipal solid waste landfill at the end of each operating day to control vectors, fires, odors, blowing litter, and scavenging. Generally, these materials must be processed so they do not allow gaps in the exposed landfill face.

Alternative Fuel   Low-polluting, nongasoline fuels such as electricity, hydrogen, propane, compressed natural gas, liquid natural gas, methanol, and ethanol.

Alternative Water Source   Nonpotable water from other than public utilities, on-site surface sources, and subsurface natural freshwater sources. Examples include gray water, on-site reclaimed water, collected rainwater, captured condensate, and rejected water from reverse osmosis systems according to the International Green Construction Code (IgCC).

American Council for an Energy-Efficient Economy (ACEEE)   A nonprofit 501 organization founded in 1980. Its mission is to act as a catalyst to advance energy efficiency policies, programs, technologies, investments, and behaviors in order to help achieve greater economic prosperity and environmental protection.

American National Standards Institute (ANSI)   A private nonprofit organization that oversees the development of voluntary consensus standards for products, services, processes, systems, and personnel in the United States. ANSI is also actively engaged in accrediting programs that assess conformance to standards, including globally recognized cross-sector programs such as the ISO (International Organization for Standardization) 9000 (quality) and ISO 4000 (environmental) management systems.

American Society of Heating, Refrigerating, and Air-Conditioning Engineers (ASHRAE)   ASHRAE's members comprise building services engineers, architects, mechanical contractors, building owners, equipment manufacturers' employees, and others concerned with the design and construction of Heating, Ventilation, Air-Conditioning, and Refrigeration (HVAC&R) systems in buildings.

ASHRAE 55-2010   Establishes the ranges of indoor environmental conditions to achieve acceptable thermal comfort for building occupants.

ASHRAE 621-2010   Sets the minimum ventilation requirements for commercial and institutional buildings to ensure optimal indoor air quality and minimize occupant negative health effects.

ASHRAE 901-2010   Covers many aspects of energy management, such as total consumption, HVAC and lighting systems, and individual circuits.

Annual Sunlight Exposure (ASE)   A metric that describes the potential for visual discomfort in interior work environments. It is defined as the percentage of an analysis area that exceeds a specified direct sunlight illuminance level more than a specified number of hours per year.

Appurtenance   A built-in, nonstructural portion of a roof system. Examples include skylights, ventilators, mechanical equipment, partitions, and solar energy panels.

Area Median Income   Midpoint in the family income range for a metropolitan statistical area, the nonmetro parts of a region, or local equivalent to either. The figure often is used as a basis to stratify incomes into low, moderate, and upper ranges.

ASE1000,250 It reports the percentage of sensors in the analysis area, using a maximum 2-foot spacing between points, that are found to be exposed to more than 1000 lux of direct sunlight for more than 250 hours per year, before any operable blinds or shades are deployed to block sunlight, considering the same 10-hour/day analysis period as sDA and using comparable simulation methods.

Assembly A product formulated from multiple materials (e.g., concrete) or a product made up of subcomponents (e.g., a workstation).

Attendance Boundary The limits used by school districts to determine what school students attend based on where they live.

Automated Dynamic Façade Systems Daylighting control devices whose position or light transmission level can be automatically changed by a control system to address sunlight penetration or perceived glare in the space. Acceptable automated dynamic façade systems include interior automated window blinds or shades; exterior automated louvers, shades, or blinds; or automatically controlled dynamic glazing. Automated methods of sunlight penetration or perceived glare control do not include manually operated interior or exterior façade shading systems; manually operated dynamic glazing; or fixed exterior overhangs, fins, shades, screens, awnings, or louvers whose position on the fenestration cannot be automatically changed or adjusted. Automated dynamic façade systems are allowed to have manual override but must default back to automated operation after a predefined period of no longer than 2 hours. Dynamic glazing is further defined in ASHRAE Standard 90.1 and the International Energy Conservation Code (IECC).

Average LED Intensity (ALI) The illumination output for light-emitting diode lamps, as specified in the International Commission on Illumination Standard 127–2007.

Average Rated Flush Volume For tank-type toilets with dual-flush capabilities, the average rated flush volume is the average of one full flush and two reduced flushes as specified by the manufacturer.

B

Base Building Materials and products that make up the building or are permanently and semipermanently installed in the project (e.g., flooring, casework, wall coverings).

Baseline Building Performance The annual energy cost for a building design, used as a baseline for comparison with above-standard design.

Baseline Condition Before the LEED project was initiated, but not necessarily before any development or disturbance took place. Baseline conditions describe the state of the project site on the date the developer acquired rights to a majority of its buildable land through purchase or option to purchase.

Baseline Water Consumption A calculated projection of building water use assuming code-compliant fixtures and fittings with no additional savings compared with the design case or actual water meter data.

Basis of Design (BOD) The information necessary to accomplish the owner's project requirements, including system descriptions, indoor environmental quality criteria, design assumptions, and references to applicable codes, standards, regulations, and guidelines.

Bicycle Network   A continuous network consisting of any combination of the following: (1) off-street bicycle paths or trails at least 8 feet (2.5 meters) wide for a two-way path and at least 5 feet (1.5 meters) wide for a one-way path; (2) physically designated on-street bicycle lanes at least 5 feet (1.5 meters) wide; (3) streets designed for a target speed of 25 miles per hour (40 kilometers/hour) or less.

Bicycling Distance   The distance that a bicyclist must travel between origins and destinations, the entirety of which must be on a bicycle network.

Bio-Based Material   Commercial or industrial products (other than food or feed) that are composed in whole, or in significant part, of biological products; renewable agricultural materials (including plant, animal, and marine materials); or forestry materials. For the purposes of LEED, this excludes leather and other animal hides.

Black Water   Wastewater containing urine or fecal matter that should be discharged to the sanitary drainage system of the building or premises in accordance with the International Plumbing Code. Wastewater from kitchen sinks (sometimes differentiated by the use of a garbage disposal), showers, or bathtubs is considered black water under some state or local codes.

Block Length   The distance along a block face; specifically, the distance from an intersecting right-of-way (ROW) edge along a block face, when that face is adjacent to a qualifying circulation network segment, to the next ROW edge intersecting that block face, except for intersecting alley ROWs.

Blowdown   The removal of makeup water from a cooling tower or evaporative condenser recirculation system to reduce concentrations of dissolved solids.

Brownfield   Real property or the expansion, redevelopment, or reuse of which may be complicated by the presence or possible presence of a hazardous substance, pollutant, or contaminant.

BUG Rating   A luminaire classification system that classifies luminaires in terms of backlight (B), uplight (U), and glare (G). BUG ratings supersede the former cutoff ratings.

Buildable Land   The portion of the site where construction can occur, including land voluntarily set aside and not constructed on. When used in density calculations, buildable land excludes public rights-of-way and land excluded from development by codified law.

Building Exterior   A structure's primary and secondary weatherproofing system, including waterproofing membranes and air- and water-resistant barrier materials and all building elements outside that system.

Building Interior   Everything inside a structure's weatherproofing membrane.

Bus Rapid Transit   An enhanced bus system that operates on exclusive bus lanes or other transit rights-of-way. The system is designed to combine the flexibility of buses with the efficiency of rail.

C

Carbon Offset   A unit of carbon dioxide equivalent that is reduced, avoided, or sequestered to compensate for emissions occurring elsewhere (World Resources Institute).

Chain of Custody (CoC)   A procedure that tracks a product from the point of harvest or extraction to its end use, including all successive stages of processing, transformation, manufacturing, and distribution.

Charrette   An intensive, multiparty workshop that brings people from different disciplines and backgrounds together to explore, generate, and collaboratively produce design options.

Chlorofluorocarbon (CFC)-Based Refrigerant   A fluid, containing hydrocarbons, that absorbs heat from a reservoir at low temperatures and rejects heat at higher temperatures. When emitted into the atmosphere, CFCs cause depletion of the stratospheric ozone layer.

Circulation Network   All motorized, nonmotorized, and mixed-mode travel ways permanently accessible to the public, not including driveways, parking lots, highway access ramps, and rights-of-way exclusively dedicated to rail. It is measured in linear feet.

Civil Twilight   The point in time in the morning (dawn) or evening (dusk) when the center of the sun is geometrically 6° below the horizon. Under good weather conditions, civil twilight is the best time to distinguish terrestrial objects clearly. Before civil twilight in the morning and after civil twilight in the evening, artificial illumination normally is required to carry on ordinary outdoor activities.

Classroom or Core Learning Space   A space that is regularly occupied and used for educational activities. In such space, the primary functions are teaching and learning, and good speech communication is critical to students' academic achievement. (Adapted from ANSI S12.60.)

Clean Waste   Nonhazardous materials left over from construction and demolition. Clean waste excludes lead and asbestos.

Clear Glazing   Glass that is transparent and allows a view through the fenestration. Diffused glazing allows only daylighting.

Closed-Loop Cooling   A system that acts as a heat sink for heat-rejecting building and medical equipment by recirculating water. Because the water is sealed within the system, some closed-loop cooling systems use nonpotable water (e.g., recycled process water harvested from an air handler's cooling coil condensate).

Color Rendering Index   A measurement from 0 to 100 that indicates how accurately an artificial light source, as compared with an incandescent light, displays hues. The higher the index number, the more accurately the light is rendering colors. Incandescent lighting has a color rendering index above 95; standard high-pressure sodium lighting (e.g., orange-hued roadway lights) measures approximately 25; many fluorescent sources using rare earth phosphors have a color rendering index of 80 and above. (Adapted from U.S. Energy Star.)

Combination Oven Discharge   Water released from an oven that includes a steam cycle or option.

Combined Heat and Power   An integrated system that captures the heat, otherwise unused, generated by a single fuel source in the production of electrical power. Also known as cogeneration. (Adapted from the U.S. Environmental Protection Agency.)

Commingled Waste   Building waste streams that are combined on the project site and hauled away for sorting into recyclable streams. Also known as single-stream recycling.

Commissioning (Cx)   The process of verifying and documenting that a building and all of its systems and assemblies are planned, designed, installed, tested, operated, and maintained to meet the owner's project requirements.

Commissioning Authority (CxA) The individual designated to organize, lead, and review the completion of commissioning process activities. The CxA facilitates communication among the owner, designer, and contractor to ensure that complex systems are installed and function in accordance with the owner's project requirements.

Conductivity The measurement of the level of dissolved solids in water, using the ability of an electric current to pass through water. Because it is affected by temperature, conductivity is measured at 25°C for standardization.

Construction Impact Zone The project's development footprint plus the areas around the improvement where construction crews, equipment, and/or materials are staged and moved during construction.

Conventional Irrigation A region's most common system for providing water to plants by nonnatural means. A conventional irrigation system commonly uses pressure to deliver water and distributes it through sprinkler heads above the ground.

Cooling Tower Blowdown The water discharged from a cooling tower typically because increased salinity or alkalinity has caused scaling. Cooling tower blowdown may be too saline for use in landscape irrigation.

Covenants, Conditions, and Restrictions (CC&R) Limitations that may be placed on a property and its use and are made a condition of holding title or lease.

Cradle-to-Gate Assessment Analysis of a product's partial life cycle, from resource extraction (cradle) to the factory gate (before it is transported for distribution and sale). It omits the use and the disposal phases of the product.

Cul-de-Sac A segment of the circulation network that terminates without intersecting another segment of the circulation network.

Cultural Landscape An officially designated geographic area that includes both cultural and natural resources associated with a historic event, activity, or person or that exhibits other significant cultural or aesthetic values.

Current Facilities Requirements (CFRs) The implementation of the owner's project requirements, developed to confirm the owner's current operational needs and requirements.

Customer For the purposes of PEER, this refers to the consumer served by the grid operator. For city projects, this includes customers of all classes billed by the utility, municipality, or third-party supplier. For campuses, this includes business and groups of individuals (i.e., building tenants or university departments) that are served by the grid operator. In the case where the building owner is the grid operator, "customer" refers to the individuals or groups of individuals on the consumer side of the meter, regardless of billing practices.

D

Dedicated Storage A designated area in a building space or a central facility that is sized and allocated for a specific task, such as the collection of recyclable waste. Signage often indicates the type of recyclable waste stored there. Some waste streams, such as mercury-based light bulbs, sensitive paper documents, biomedical waste, or batteries, may require particular handling or disposal methods. Consult the municipality's safe storage and disposal procedures or use guidelines posted on the U.S. Environmental Protection Agency website at www.epa.gov.

Demand Response (DR)   A change in electricity use by demand-side resources from their normal consumption patterns in response to changes in the price of electricity or to incentive payments designed to induce lower electricity use at times of high wholesale market prices or when system reliability is jeopardized.

Demand Response (DR) Event   A specific period of time when the utility or independent service operator calls for a change in the pattern or level of use in grid-based electricity from its program participants. Also known as a curtailment event.

Demountable Partition   A temporary interior wall that can be easily reconfigured. In a healthcare facility, acoustical concerns and embedded equipment, as in a surgery suite, may prevent demountable partitions from being used.

Densely Occupied Space   An area with a design occupant density of 25 people or more per 1000 square feet (93 square meters).

Density   A measure of the total building floor area or dwelling units on a parcel of land relative to the buildable land of that parcel. Units for measuring density may differ according to credit requirements. Does not include structured parking.

Departmental Gross Area (DGA)   The floor area of a diagnostic and treatment clinical department, calculated from the centerline of the walls separating the department from adjacent spaces. Walls and circulation space within the department are included in the calculation. This calculation excludes inpatient units.

Development Footprint   The total land area of a project site covered by buildings, streets, parking areas, and other typically impermeable surfaces constructed as part of the project.

Differential Durability   A state in which two materials with different life spans make up one complete component. If one material wears out and cannot be separated and replaced, the entire product must be thrown away.

Direct Access   A means of entering a space without having to leave the floor or pass through another patient's room, dedicated staff space, service or utility space, or major public space. Patients' and public circulation corridors, common sitting areas, and waiting and day space may be part of a direct access route.

Direct Sunlight   An interior horizontal measurement of 1000 lux or more of direct beam sunlight that accounts for window transmittance and angular effects and excludes the effect of any operable blinds, with no contribution from reflected light (i.e., a zero-bounce analysis) and no contribution from the diffuse sky component. (Adapted from IES.)

District Energy System (DES)   A central energy conversion plant and transmission and distribution system that provides thermal energy to a group of buildings (e.g., a central cooling plant on a university campus). It does not include central energy systems that provide only electricity.

Diverse Use   A distinct business or organization that provides goods or services intended to meet daily needs and is publicly available. Automated facilities such as ATMs or vending machines are not included.

Downstream Equipment   The heating and cooling systems, equipment, and controls located in the project building or on the project site and associated with transporting the thermal energy of the district energy system (DES) into heated and cooled spaces. Downstream equipment includes the thermal connection or interface with the DES, secondary distribution systems in the building, and terminal units. Drift water droplets are carried from a cooling tower or evaporative condenser by a stream of air passing through the system. Drift eliminators capture these droplets and return them to the reservoir at the bottom of the cooling tower or evaporative condenser for recirculation.

Drift   Water droplets carried from a cooling tower or evaporative condenser by a stream of air passing through the system. Drift eliminators capture these droplets and return them to the reservoir at the bottom of the cooling tower or evaporative condenser for recirculation.

Durable Goods   Products with a useful life of approximately 2 years or more and that are replaced infrequently. Examples include furniture, office equipment, appliances, external power adapters, televisions, and audiovisual equipment.

Durable Goods Waste Stream   The flow of long-lasting products from the project building after they are fully depreciated and have reached the end of their useful life for normal business operations. It includes leased durable goods returned to their owner but does not include durable goods that remain functional and are moved to another floor or building.

E

Electric Vehicle Supply Equipment   The conductors, including the ungrounded, grounded, and equipment grounding conductors; the electric vehicle connectors; attachment plugs; and all other fittings, devices, power outlets or apparatuses installed specifically for the purpose of delivering energy from the premises wiring to the electric vehicle. (National Electric Codes and California Article 625.)

Electronic Waste   Discarded office equipment (computers, monitors, copiers, printers, scanners, fax machines); appliances (refrigerators, dishwashers, water coolers); external power adapters; and televisions and other audiovisual equipment.

Elemental Mercury   Mercury in its purest form (rather than a mercury-containing compound), the vapor of which is commonly used in fluorescent and other bulb types.

Emergency Lighting   A luminaire that operates only during emergency conditions and is always off during normal building operation.

Employment Center   A nonresidential area of at least 5 acres (2 hectares) with a job density of at least 50 employees per net acre (at least 125 employees per hectare net).

Enclosure   The exterior plus semiexterior portions of the building. Exterior consists of the elements of a building that separate conditioned spaces from the outside (i.e., the wall assembly). Semiexterior consists of the elements of a building that separate conditioned space from unconditioned space or that encloses semiheated space through which thermal energy may be transferred to or from the exterior or conditioned or unconditioned spaces (e.g., attic, crawl space, basement).

Energy Service Provider   A designation that allows an outside entity, such as USGBC, to access water and energy usage information that a building management team maintains with Energy Star Portfolio Manager or a similar tool.

Engineered Nanomaterial   A substance designed at the molecular (nanometer) level. Because of its small size, it has novel properties generally not seen in its conventional bulk counterpart. See the Australian National Industrial Chemicals Notification and Assessment Scheme, https://chemycal.com/dap/files/Guidance/201408-NICNAS.pdf.

Environmental Product Declaration   A statement that the item meets the environmental requirements of ISO 14021–1999, ISO 14025–2006 and EN 15804, or ISO 21930–2007.

Evapotranspiration   The combination of evaporation and plant transpiration into the atmosphere. Evaporation occurs when liquid water from soil, plant surfaces, or water bodies becomes vapor. Transpiration is the movement of water through a plant and the subsequent loss of water vapor.

Extended Producer Responsibility   Measures undertaken by the maker of a product to accept its own and sometimes other manufacturers' products as postconsumer waste at the end of the products' useful life. Producers recover and recycle the materials for use in new products of the same type. To count toward credit compliance, a program must be widely available. For carpet, extended producer responsibility must be consistent with NSF/ANSI 140–2007. Also known as closed-loop program or product take-back.

Extensive Vegetated Roof   A roof that is covered with plants and typically not designed for general access. Usually an extensive system is a rugged green roof that requires little maintenance once established. The planting medium in extensive vegetated roofs ranges from 1 to 6 inches in depth. (Adapted from the U.S. Environmental Protection Agency.) The exterior vegetated surface area is the total area of vegetation on the project site, including vegetated roofs and turf grass.

External Meter   A device installed on the outside of a water pipe to record the volume of water passing through it. Also known as a clamp-on meter.

F

Floor-Area Ratio (FAR)   The density of nonresidential land use, exclusive of parking, measured as the total nonresidential building floor area divided by the total buildable land area available for nonresidential structures. For example, on a site with 10,000 square feet (930 square meters) of buildable land area, an FAR of 1.0 would be 10,000 square feet (930 square meters) of building floor area. On the same site, an FAR of 1.5 would be 15,000 square feet (1395 square meters), an FAR of 2.0 would be 20,000 square feet (1860 square meters), and an FAR of 0.5 would be 5000 square feet (465 square meters).

Foundation Drain   The water discharged from a subsurface drainage system. If a building foundation is below the water table, a sump pump may be required. Discharge from the sump may be stored and used for irrigation.

Freight Village   A cluster of freight-related businesses that include intermodal transfer operations. Freight villages may offer logistics services, integrated distribution, warehousing capabilities, showrooms, and support services. Such support services may include security, maintenance, mail, banking, customs and import management assistance, cafeterias, restaurants, office space, conference rooms, hotels, and public or activity center transportation.

Functional Entry   A building opening designed to be used by pedestrians and open during regular business hours. It does not include any door exclusively designated as an emergency exit or a garage door not designed as a pedestrian entrance.

Furniture and Furnishings   The stand-alone furniture items purchased for the project, including individual and group seating; open-plan and private-office workstations; desks and tables; storage units, credenzas, bookshelves, filing cabinets, and other case goods; wall-mounted visual display products (e.g., marker boards and tack boards, excluding electronic displays); and miscellaneous items, such as easels, mobile carts, freestanding screens, installed fabrics, and movable partitions. Hospitality furniture is included as applicable to the project. Office accessories, such as desktop blotters, trays, tape dispensers, waste baskets, and all electrical items, such as lighting and small appliances, are excluded.

G

Grams per Brake Horsepower Hour   Metric used to communicate how many grams of emissions (e.g., nitrogen oxide or particulate matter) are emitted by an engine of a specific horsepower rating over a 1-hour period.

Gray Water   "Untreated household wastewater which has not come into contact with toilet waste. Graywater includes used water from bathtubs, showers, bathroom wash basins, and water from clothes-washers and laundry tubs. It must not include wastewater from kitchen sinks or dishwashers" (Uniform Plumbing Code, "Appendix G, Gray Water Systems for Single-Family Dwellings"); "wastewater discharged from lavatories, bathtubs, showers, clothes washers and laundry sinks" (International Plumbing Code, "Appendix C, Gray Water Recycling Systems"). Some states and local authorities allow kitchen sink wastewater to be included in graywater. Other differences can likely be found in state and local codes. Project teams should comply with the graywater definition established by the authority having jurisdiction in the project area.

Green Infrastructure   A soil- and vegetation-based approach to wet weather management that is cost-effective, sustainable, and environmentally friendly. Green infrastructure management approaches and technologies infiltrate, evapotranspire, capture, and reuse stormwater to maintain or restore natural hydrologies. (Adapted from the U.S. Environmental Protection Agency.)

Green Power   A subset of renewable energy composed of grid-based electricity produced from renewable energy sources.

Green Vehicles   Vehicles achieving a minimum green score of 45 on the American Council for an Energy Efficient Economy (ACEEE) annual vehicle rating guide (or a local equivalent for projects outside the United States).

Greenfield   Area that has not been graded, compacted, cleared, or disturbed and that supports (or could support) open space, habitat, or natural hydrology.

Gross Bulk Generation   The sum of bulk purchased electricity delivered to the site. Bulk generation delivered to the site should be adjusted for transmission and distribution losses to the site.

Gross Floor Area (GFA)   Generally, the gross floor area is the sum of the floor areas of the spaces within the building, including basements, mezzanine and intermediate-floor tiers, and penthouses with headroom height of 7.5 feet (2.2 meters) or greater. Measurements must be taken from the exterior faces of exterior walls OR from the centerline of walls separating buildings OR (for LEED ID+C projects) from the centerline of walls separating spaces. Excludes nonenclosed (or nonenclosable) roofed-over areas, such as exterior covered walkways, porches, terraces or steps, roof overhangs, and similar features. Excludes air shafts, pipe trenches, chimneys, and floor area dedicated to the parking and circulation of motor vehicles.

Gross Local Generation   The sum of electric power generated at the generator terminals of local generators. This does not include any hotel loads producing electric power in the plant.

H

Hardscape   The inanimate elements of the building landscaping. It includes pavement, roadways, stone walls, wood and synthetic decking, concrete paths and sidewalks, and concrete, brick, and tile patios.

Hazardous Material   Any item or agent (biological, chemical, physical) that has the potential to cause harm to humans, animals, or the environment, either by itself or through interaction with other factors.

Heat Island Effect   The thermal absorption by hardscape, such as dark, nonreflective pavement and buildings, and its subsequent radiation to surrounding areas. Other contributing factors may include vehicle exhaust, air conditioners, and street equipment. Tall buildings and narrow streets reduce airflow and exacerbate the effect.

Highway   A transportation thoroughfare intended for motor vehicles with limited access points, prohibitions on human-powered vehicles, and higher speeds than local roads. A highway generally connects cities and towns.

Historic Building   A building or structure with historic, architectural, engineering, archeological, or cultural significance that is listed or determined to be eligible as a historic structure or building or as a contributing building or structure in a designated historic district. The historic designation must be made by a local historic preservation review board or similar body, and the structure must be listed in a state register of historic places, be listed in the National Register of Historic Places (or a local equivalent outside the United States), or have been determined eligible for listing.

Historic District   A group of buildings, structures, objects, and sites that have been designated or determined to be eligible as historically and architecturally significant and categorized as either contributing or noncontributing to the historic nature of the district.

Homogeneous Material   An item that consists of only one material throughout or a combination of multiple materials that cannot be mechanically disjointed, excluding surface coatings.

Hot-Mopped Asphalt for Roofing   A mopping-grade oxidized asphalt as defined by the National Institute for Occupational Safety and Health (NIOSH): "An oxidized asphalt used principally in the construction of built-up roofing (BUR) and some modified bitumen systems; mopping-grade asphalts are produced in four grades (Types I through IV), according to the steepness of the roof."

Hydrozone   A group of plantings with similar water needs.

I

Illuminance   The incident luminous flux density on a differential element of surface located at a point and oriented in a particular direction, expressed in lumens per unit area. Since the area involved is differential, it is customary to refer to this as illuminance at a point. The unit name depends on the unit of measurement for area: foot-candles if square feet are used for area, and lux if square meters are used. (Adapted from IES.) In lay terms, illuminance is a measurement of light striking a surface. It is expressed in foot-candles in the United States (based on square feet) and in lux in most other countries (based on square meters).

Impervious Surface   An area of ground that development and building have modified in such a way that precipitation cannot infiltrate downward through the soil. Examples of impervious surfaces include roofs, paved roads and parking areas, sidewalks, and soils that have been compacted either by design or by use.

Individual Occupant Space   An area where an occupant performs distinct tasks. Individual occupant spaces may be within multioccupant spaces and should be treated separately where possible.

Industrial Process Water Any water discharged from a factory setting. Before this water can be used for irrigation, its quality needs to be checked. Saline or corrosive water should not be used for irrigation.

Infill Site A site where at least 75 percent of the land area, exclusive of rights of way, within ½ mile (800 meters) of the project boundary is previously developed. A street or other right of way does not constitute previously developed land; it is the status of property on the other side of right of way or the street that matters.

Infiltration The heating, ventilating, and air-conditioning (HVAC) uncontrolled inward air leakage to conditioned spaces through unintentional openings in ceilings, floors, and walls from unconditioned spaces or the outdoors caused by the same pressure differences that induce exfiltration (ASHRAE 62.1–2010).

Informal Transit A transit service in which individuals travel together in a passenger car or small van that seats at least four people. It can include human-powered conveyances, which must accommodate at least two people. It must include an enclosed passenger seating area, fixed route service, fixed fare structure, regular operation, and the ability to pick up multiple riders.

Infrared (Thermal) Emittance A value between 0 and 1 (or 0 percent and 100 percent) that indicates the ability of a material to shed infrared radiation (heat). A cool roof should have a high thermal emittance. The wavelength range for radiant energy is roughly 5 to 40 micrometers. Most building materials (including glass) are opaque in this part of the spectrum and have an emittance of roughly 0.9, or 90 percent. Clean, bare metals, such as untarnished galvanized steel, have a low emittance and are the most important exceptions to the 0.9 rule. In contrast, aluminum roof coatings have intermediate emittance levels. (Adapted from Lawrence Berkeley National Laboratory.)

Inpatient An individual admitted to a medical, surgical, maternity, specialty, or intensive-care unit for a length of stay exceeding 23 hours.

Inpatient Unit Any medical, surgical, maternity, specialty, or intensive-care unit where an individual receives care for more than 23 hours.

Integral Labeling An information conveyance system that cannot be easily removed. For furniture, such labeling may include radio-frequency identification, engraving, embossing, or other permanent marking containing information on material origin, properties, and date of manufacture.

Integrated Pest Management A method of pest management that protects human health and the surrounding environment and improves economic returns through the most effective, least risk option.

Integrated Project Delivery An approach that involves people, systems, and business structures (contractual and legal agreements) and practices. The process harnesses the talents and insights of all participants to improve results, increase value to the owner, reduce waste, and maximize efficiency through all phases of design, fabrication, and construction. (Adapted from the American Institute of Architects.)

Intensive Vegetated Roof A roof that, compared with an extensive vegetated roof, has greater soil volume, supports a wider variety of plants (including shrubs and trees), and allows a wider variety of uses (including human access). The depth of the growing medium is an important factor in determining habitat value. The native or adapted plants selected for the roof should support the site's endemic wildlife populations. (Adapted from Green Roofs for Healthy Cities.)

Interior Floor Finish   All the layers applied over a finished subfloor or stairs, including stair treads and risers, ramps, and other walking surfaces. Interior finish excludes building structural members, such as beams, trusses, studs, subfloors, or similar items. Interior finish also excludes nonfull-spread wet coatings or adhesives.

Interior Wall and Ceiling Finish   All the layers comprising the exposed interior surfaces of buildings, including fixed walls, fixed partitions, columns, exposed ceilings, and interior wainscoting, paneling, interior trim or other finish applied mechanically or for decoration, acoustical correction, surface fire resistance, or similar purposes.

Intermodal Facility   A venue for the movement of goods in a single loading unit or road vehicle that uses successively two or more modes of transportation without the need to handle the goods themselves.

Interstitial Space   An intermediate space located between floors with a walk-on deck, often used to run the majority of the utility distribution and terminal equipment, thus permitting convenient installation, maintenance, and future modifications.

Invasive Plant   Nonnative vegetation that has been introduced to an area and that aggressively adapts and reproduces. The plant's vigor combined with a lack of natural enemies often leads to outbreak populations. (Adapted from the U.S. Department of Agriculture.)

Information Technology (IT) Annual Energy   Electricity consumption by information technology and telecom equipment, which includes servers, networking, and storage equipment, over the course of a year.

L

Lamp   A device emitting light in a fixture, excluding lamp housing and ballasts. Light-emitting diodes packaged as traditional lamps also meet this definition.

Lamp Life   The useful span of operation of a source of artificial light, such as bulbs. Lamp life for fluorescent lights is determined by testing 3 hours on for every 20 minutes off. For high-density discharge lamps, the test is based on 11 hours on for every 20 minutes off. Lamp life depends on whether the start ballast is programmed or instant. This information is published in the manufacturers' information. Also known as rated average life.

Land Trust   A private, nonprofit organization that, as all or part of its mission, actively works to conserve land by undertaking or assisting in conservation easement or land acquisition, or by its stewardship of such land or easements. (Adapted from the Land Trust Alliance.)

Land-Clearing Debris and Soil   Materials that are natural (e.g., rock, soil, stone, vegetation). Materials that are man-made (e.g., concrete, brick, cement) are considered construction waste even if they were on site.

Landscape Water Requirement (LWR)   The amount of water that the site landscape area(s) requires for the site's peak watering month.

Lead Free   A label, defined by U.S. Environmental Protection Agency regulations under the Safe Drinking Water Act, that allows small amounts of lead in solders, flux, pipes, pipe fittings, and well pumps.

Least-Risk Pesticide   A registered pesticide in the Tier III (lowest toxicity) category, using the San Francisco Hazard Ranking system, or a pesticide that meets the requirements in the San Francisco Pesticide Hazard Screening Protocol and is sold as a self-contained bait or as a crack-and-crevice treatment used in areas inaccessible to building occupants. Rodenticides are never considered least-risk pesticides.

LEED ND The portion of the site where construction can occur, including land voluntarily set aside and not constructed on. When used in density calculations, buildable land excludes public rights of way and land excluded from development by codified law or LEED for Neighborhood Development prerequisites.

Length of Stay The amount of time a person remains in a healthcare facility as an admitted patient.

Life-Cycle Assessment An evaluation of the environmental effects of a product from cradle to grave, as defined by ISO 14040–2006 and ISO 14044–2006.

Life-Cycle Inventory A database that defines the environmental effects (inputs and outputs) for each step in a materials or assembly's life cycle. The database is specific to countries and regions within countries.

Light Pollution Waste light from building sites that produces glare, is directed upward to the sky, or is directed off the site. Waste light does not increase nighttime safety, utility, or security and needlessly consumes energy.

Light Rail Transit service using two- or three-car trains in a right of way that is often separated from other traffic modes. Spacing between stations tends to be ½ mile or more, and maximum operating speeds are typically 40–55 miles per hour (65–90 kilometers per hour). Light-rail corridors typically extend 10 or more miles (16 kilometers).

Light Trespass Obtrusive illumination that is unwanted because of quantitative, directional, or spectral attributes. Light trespass can cause annoyance, discomfort, distraction, or loss of visibility.

Load Shedding An intentional action by a utility to reduce the load on the system. Load shedding is usually conducted during emergency periods, such as capacity shortages, system instability, or voltage control.

Long-Term Bicycle Storage Bicycle parking that is easily accessible to residents and employees and covered to protect bicycles from rain and snow.

Low-Cost Improvement An operational improvement, such as a repair, upgrade, or provided staff training or retraining. In LEED, the project team determines the reasonable upper limit for low-cost improvements based on facility resources and operating budgets.

Low-Impact Development (LID) An approach to managing rainwater runoff that emphasizes on-site natural features to protect water quality by replicating the natural land cover hydrologic regime of watersheds and addressing runoff close to its source. Examples include better site design principles (e.g., minimizing land disturbance, preserving vegetation, minimizing impervious cover), and design practices (e.g., rain gardens, vegetated swales and buffers, permeable pavement, rainwater harvesting, soil amendments). These are engineered practices that may require specialized design assistance.

M

Major Renovation Extensive alteration work in addition to work on the exterior shell of the building and/or primary structural components and/or the core and peripheral Mechanical, Electrical, Plumbing (MEP) and service systems and/or site work. Typically, the extent and nature of the work is such that the primary function space cannot be used for its intended purpose while the work is in progress and where a new certificate of occupancy is required before the work area can be reoccupied.

Makeup Water   Water that is fed into a cooling tower system or evaporative condenser to replace water lost through evaporation, drift, bleed-off, or other causes.

Manage (Rainwater) on Site   To capture and retain a specified volume of rainfall to mimic natural hydrologic function. Examples of rainwater management include strategies that involve evapotranspiration, infiltration, and capture and reuse.

Master Plan Boundary   The limits of a site master plan. The master plan boundary includes the project area and may include all associated buildings and sites outside of the LEED project boundary. The master plan boundary considers future sustainable use, expansion, and contraction.

Mean Lumen Output   A measurement of a source's emitted light derived from industry standards, taken with an instant-start ballast that has a ballast factor of 1.0 as measured at 40 percent of lamp life (except for T-5 lamps, which use a program-start ballast).

Medical Furnishing   An item of furniture designed for use in healthcare. Examples include surgical tables; procedure, supply, and mobile technology carts; lifting and transfer aids; supply closet carts and shelving; and overbed tables.

Metering Control   A regulator that limits the flow time of water, generally a manual-on and automatic-off device, most commonly installed on lavatory faucets and showers.

Mixed Paper   White and colored paper, envelopes, forms, file folders, tablets, flyers, cereal boxes, wrapping paper, catalogs, magazines, phone books, and photos.

Modular and Movable Casework   Shelving and cabinetry designed to be easily installed, moved, or reconfigured. In a retail setting, items that are movable but semipermanently attached using mechanical fastening systems for operational use are considered furniture and not base building elements (e.g., a table or display bolted to the floor or shelving attached to a wall).

Mounting Height   The distance between ground level (or the work plane) and the center of the luminaire (light fixture); the height at which a luminaire is installed.

Movable Furniture and Partitions   Items that can be moved by the users without the need of tools or assistance from special trades and facilities management.

Multitenant Complex   A site that was master planned for the development of stores, restaurants, and other businesses. Retailers may share some services and common areas.

N

NAED Code   A unique five- or six-digit number used to identify specific lamps, used by the National Association of Electrical Distributors (NAED).

Native Vegetation   A species that originates in, and is characteristic of, a particular region and ecosystem without direct or indirect human actions. Native species have evolved together with other species within the geography, hydrology, and climate of that region.

Natural Refrigerant   A compound that is not man made and is used for cooling. Such substances generally have much lower potential for atmospheric damage than manufactured chemical refrigerants. Examples include water, carbon dioxide, and ammonia.

Natural Resources Conservation Service (NRCS) Soils Delineation A U.S.-based soil survey that shows the boundaries of different soil types and special soil features on the site.

Natural Site Hydrology The natural land cover function of water occurrence, distribution, movement, and balance.

Net Usable Program Area The sum of all interior areas in the project available to house the project's program. It does not include areas for building equipment, vertical circulation, or structural components.

Noninpatient Area A public space, diagnostic or treatment area, ambulatory unit, or any other space in a healthcare facility that is not for individuals who have been admitted for care.

Nonregularly Occupied Space An area that people pass through or an area used for focused activities an average of less than 1 hour per person per day. The 1-hour time frame is continuous and should be based on the time a typical occupant uses the space. For spaces that are not used daily, the 1-hour time frame should be based on the time a typical occupant spends in the space when it is in use.

Nonpotable Water Water that does not meet drinking water standards.

Nonwater Toilet Systems Dry plumbing fixtures and fittings that contain and treat human waste via microbiological processes.

Nonwater Urinal A plumbing fixture having a water flush with a trap that contains a layer of buoyant liquid that floats above the urine, blocking sewer gas and odors.

O

Occupant Control A system or switch that a person in the space can directly access and use. Examples include a task light, an open switch, and blinds. A temperature sensor, photo sensor, or centrally controlled system is not occupant controlled.

Occupiable Space An enclosed space intended for human activities, excluding those spaces that are intended primarily for other purposes, such as storage rooms and equipment rooms, and that are occupied only occasionally and for short periods of time (ASHRAE 62.1–2010).

Occupied Space Enclosed space intended for human activities, excluding those spaces that are intended primarily for other purposes, such as storage rooms and equipment rooms, and that are only occupied occasionally and for short periods of time. Occupied spaces are further classified as regularly occupied or nonregularly occupied spaces based on the duration of the occupancy, individual or multioccupant based on the quantity of occupants, and densely or nondensely occupied spaces based on the concentration of occupants in the space.

On-Site Wastewater Treatment The transport, storage, treatment, and disposal of wastewater generated on the project site.

Ongoing Consumable A product that has a low cost per unit and is regularly used and replaced in the course of business. Examples include paper, toner cartridges, binders, batteries, and desk accessories. Also known as ongoing purchases.

Open-Grid Pavement System Pavements that consist of loose substrates supported by a grid of a more structurally sound grid or webbing. Pervious concrete and porous asphalt are not considered open grid as they are considered bounded materials. Unbounded, loose substrates do not transfer and store heat like bound and compacted materials do.

Openable Window Area   The free, unobstructed area through the opening.

Operations and Maintenance (O&M) Plan   A plan that specifies major system operating parameters and limits, maintenance procedures and schedules, and documentation methods necessary to demonstrate proper operation and maintenance of an approved emissions control device or system.

Ornamental Luminaire   A luminaire intended for illuminating portions of the circulation network that also serves an ornamental function, in addition to providing optics that effectively deliver street lighting, and has a decorative or historical period appearance.

Outpatient   A patient who is not hospitalized for 24 hours or more but who visits a hospital, clinic, or associated healthcare facility for diagnosis or treatment.

Owner's Project Requirements (OPR)   A written document that details the ideas, concepts, and criteria determined by the owner to be important to the success of the project.

P

Patient Position   A patient bed, infusion chair, recovery room bay, or other location where a patient receives clinical care.

Peak Demand   The maximum electricity load at a specific point in time or over a period of time.

Peak Watering Month   The month with the greatest deficit between evapotranspiration and rainfall. This is the month when the plants in the site's region potentially require the most supplemental water, typically a midsummer month.

Performance Excellence in Electricity Renewal (PEER)   According to USGBC, it is the nation's first comprehensive, consumer-centric, data-driven system for evaluating power system performance; it's also modeled after LEED. The PEER rating system helps fill a major gap in the smart grid movement, providing an opportunity for power systems to gain competitive advantage by differentiating their performances and demonstrating meaningful outcomes. In the process, the metrics serve as a tool to accelerate transformation in the marketplace.

How PEER works? By looking at power systems across four outcome categories PEER allows all stakeholders to both evaluate and set the standard for system performance that best meets customers' needs.

- Enabling customer action: The intent of this category is to assess customer participation as a resource in grid improvements and enable private investment and innovation.
- Operational efficiency: The intent of this category is to assess spending practices and assist in the identification and elimination of wasteful spending through performance benchmarks.
- Reliability, power quality, and safety: The intent of this category is to assess the quality of power delivery and reduce negative impacts on the customer.
- Energy efficiency and environment: The intent of this category is to assess the environmental impact of electricity generation and transmission and encourage the adoption of clean energy.

Permanent Interior Obstruction   A structure that cannot be moved by the user without tools or assistance from special trades and facilities management. Examples include lab hoods, fixed partitions, demountable opaque full- or partial-height partitions, some displays, and equipment.

Permanent Peak Load Shifting   The transfer of energy consumption to off-peak hours, when demand for power is lower and energy is therefore less expensive.

Permeable Pavement   A paved surface that allows water runoff to infiltrate into the ground.

Persistent Bioaccumulative Toxic Chemical   A substance that poses a long-term risk to both humans and the environment because it remains in the environment for long periods, increases in concentration as it moves up the food chain, and can travel far from the source of contamination. Often these substances can become more potent and harmful to ecosystems the longer they persist.

Place of Respite   An area that connects healthcare patients, visitors, and staff to health benefits of the natural environment. (Adapted from "Green Guide for Health Care Places of Respite Technical Brief.")

Plug Load or Receptacle Load   The electrical current drawn by all equipment connected to the electrical system via a wall outlet.

Postconsumer Recycled Content   Waste generated by households or commercial, industrial, and institutional facilities in their role as end users of a product that can no longer be used for its intended purpose.

Potable Water   Water that meets or exceeds the U.S. Environmental Protection Agency drinking water quality standards (or a local equivalent outside the United States) and is approved for human consumption by the state or local authorities having jurisdiction; it may be supplied from wells or municipal water systems.

Power Distribution Unit Output   The electrical power from a device that allocates power to and serves information technology (IT) equipment. Power distribution unit (PDU) output does not include efficiency losses of any transformation that occurs within the PDU, but it can include downstream non-IT ancillary devices installed in IT racks, such as fans. If the PDU system supports non-IT equipment (e.g., computer room air-conditioning units, computer room air handlers, in-row coolers), this equipment must be metered and subtracted from the PDU output reading. The metering approach should be consistent with the metering required for the power usage efficiency (PUE) category (e.g., continuous consumption metering for PUE Categories 1, 2, and 3).

Power Utilization Effectiveness (PUE)   A measure of how efficiently a data center uses its power; specifically, how much power is used by computing equipment rather than for cooling and other overhead.

Powered Floor Maintenance Equipment   Electric and battery-powered floor buffers and burnishers. It does not include equipment used in wet applications.

Preconsumer Recycled Content   Matter diverted from the waste stream during the manufacturing process, determined as the percentage of material, by weight. Examples include planer shavings, sawdust, bagasse, walnut shells, culls, trimmed materials, overissue publications, and obsolete inventories. The designation excludes rework, regrind, or scrap materials capable of being reclaimed within the same process that generated them (ISO 14021). Formerly known as postindustrial content.

Preferred Parking   The parking spots closest to the main entrance of a building (exclusive of spaces designated for handicapped persons). For employee parking, it refers to the spots that are closest to the entrance used by employees.

Premature Obsolescence   The wearing out or disuse of components or materials whose service life exceeds their design life. For example, a material with a potential life of 30 years is intentionally designed to last only 15 years, such that its remaining 15 years of service is potentially wasted. In contrast, components whose service life is the same as their expected use are utilized to their maximum potential.

Previously Developed   Altered by paving, construction, and/or land use that would typically have required regulatory permitting to have been initiated (alterations may exist now or in the past). Land that is not previously developed and landscapes altered by current or historical clearing or filling, agricultural or forestry use, or preserved natural area use are considered undeveloped land. The date of previous development permit issuance constitutes the date of previous development but permit issuance in itself does not constitute previous development.

Previously Developed Site   A site that, prior to the project, consisted of at least 75 percent previously developed land.

Previously Disturbed   Areas that have been graded, compacted, cleared, previously developed, or disturbed in any way. These are areas that do not qualify as "greenfield." An area where the plant community has been severely disturbed and has not recovered or the natural (native) plant biota is nearly gone.

Prime Farmland   Land that has the best combination of physical and chemical characteristics for producing food, feed, forage, fiber, and oilseed crops and that is available for these uses, as determined by the U.S. Department of Agriculture's Natural Resources Conservation Service (a U.S.-based methodology that sets criteria for highly productive soil). For a complete description of what qualifies as prime farmland, see U.S. Code of Federal Regulations, Title 7, Volume 6, Parts 400 to 699, Section 657.5.

Private Meter   A device that measures water flow and is installed downstream from the public water supply meter or as part of an on-site water system maintained by the building management team.

Process Energy   Power resources consumed in support of a manufacturing, industrial, or commercial process other than conditioning spaces and maintaining comfort and amenities for building occupants. It may include refrigeration equipment, cooking and food preparation, clothes washing, and other major support appliances.

Process Load or Unregulated Load   The load on a building resulting from the consumption or release of process energy. (ASHRAE.)

Process Water   Water that is used for industrial processes and building systems, such as cooling towers, boilers, and chillers. It can also refer to water used in operational processes, such as dishwashing, clothes washing, and ice making.

Product (Permanently Installed Building Product)   An item that arrives on the project site as either a finished element ready for installation or a component to another item assembled on site. The product unit is defined by the functional requirement for use in the project; this includes the physical components and services needed to serve the intended function of the permanently installed building product. In addition, in a similar product within a specification, each contributes as a separate product.

Project Electrical Load The project load includes electric power delivered to metered customers, buildings, or loads within the project boundary, including the electric power required to centrally produce other energy delivered to customers, buildings, or loads within the project boundary such as chiller loads. This does not include parasitic loads such as transmission and distribution losses or reactive power compensation.

Public Water Supply (PWS) A system for the provision to the public of water for human consumption through pipes or other constructed conveyances. To be considered public, such system must have at least 15 service connections or regularly serve at least 25 individuals. (Adapted from the U.S. Environmental Protection Agency.)

R

Rainwater Harvesting The capture, diversion, and storage of rain for future beneficial use. Typically, a rain barrel or cistern stores the water; other components include the catchment surface and conveyance system. The harvested rainwater can be used for irrigation.

Raw Material The basic substance from which products are made, such as concrete, glass, gypsum, masonry, metals, recycled materials (e.g., plastics and metals), oil (petroleum, polylactic acid), stone, agrifiber, bamboo, and wood.

Reclaimed Water Wastewater that has been treated and purified for reuse.

Recycled Content Defined in accordance with the International Organization of Standardization document "ISO 14021Environmental Labels and Declarations—Self-Declared Environmental Claims (Type II Environmental Labeling)."

Reference Evapotranspiration Rate The amount of water lost from a specific vegetated surface with no moisture limitation. Turf grass with a height of 120 millimeters is the reference vegetation.

Reference Soil A soil native to the project site, as described in Natural Resources Conservation Service soil surveys (or a local equivalent survey outside the United States), or undisturbed native soils within the project's region that have native vegetation, topography, and soil textures similar to the project site. For project sites with no existing soil, reference soils are defined as undisturbed native soils within the project's region that support appropriate native plant species similar to those intended for the new project.

Refurbished Material An item that has completed its life cycle and is prepared for reuse without substantial alteration of its form. Refurbishing involves renovating, repairing, restoring, or generally improving the appearance, performance, quality, functionality, or value of a product.

Regularly Occupied Space An area where one or more individuals normally spends time (more than 1 hour per person per day on average) seated or standing as they work, study, or perform other focused activities inside a building. The 1-hour time frame is continuous and should be based on the time a typical occupant uses the space. For spaces that are not used daily, the 1-hour time frame should be based on the time a typical occupant spends in the space when it is in use.

Regularly Used Exterior Entrance A frequently used means of gaining access to a building. Examples include the main building entrance as well as any building entryways attached to parking structures, underground parking garages, underground pathways, or outside spaces. Atypical entrances, emergency exits, atriums, connections between concourses, and interior spaces are not included.

Regulated Load Any building end use that has either a mandatory or a prescriptive requirement in ANSI/ASHRAE/IES Standard 90.1–2010.

Remanufactured Product An item that has been repaired or adjusted and returned to service. A remanufactured product can be expected to perform as if it were new.

Renewable Energy Energy sources that are not depleted by use. Examples include energy from the sun, wind, and small (low-impact) hydropower, plus geothermal energy and wave and tidal systems.

Renewable Energy Credit (REC) A tradable commodity representing proof that a unit of electricity was generated from a renewable resource. RECs are sold separately from electricity itself and thus allow the purchase of green power by a user of conventionally generated electricity.

Reuse Materials: The reemployment of materials in the same or a related capacity as their original application, thus extending the lifetime of materials that would otherwise be discarded. Reuse includes the recovery and reemployment of materials recovered from existing building or construction sites. Also known as salvage. Water: The practice of using water that has already been used. The terms reclaimed water, reused water, and recycled water are used interchangeably in the water industry.

Reused Area The total area of the building structure, core, and envelope that existed in the prior condition and remains in the completed design.

Revenue-Grade Meter A measurement tool designed to meet strict accuracy standards required by code or law. Utility meters are often called revenue grade because their measurement directly results in a charge to the customer.

S

Salvaged Material A construction component recovered from existing buildings or construction sites and reused. Common salvaged materials include structural beams and posts, flooring, doors, cabinetry, brick, and decorative items.

School Authority The authority responsible for decision-making about school operations, districts, personnel, financing, and future development. Examples include school boards, local governments, and religious institutions.

Scope 1 Emissions Direct greenhouse gas emissions from sources owned or controlled by the entity, such as emissions from fossil fuels burned on site. Electricity produced on site through the burning of fossil fuels is measured by the Scope 1 emissions associated with that fossil fuel.

Scope 2 Emissions Indirect greenhouse gas emissions associated with the generation of purchased electricity, heating/cooling, or steam off site, through a utility provider for the entity's consumption.

Server Input The information technology (IT) load as measured at the point of connection (e.g., power receptacle) of the IT device to the electrical power system. Server input captures the actual power load of the IT device exclusive of any power distribution losses and non-IT loads (e.g., rack-mounted fans).

Service Life The assumed length of time that a building, product, or assembly will be operational for the purposes of a life-cycle assessment.

Shared Multioccupant Space A place of congregation or where occupants pursue overlapping or collaborative tasks.

Shell Space An area designed to be fitted out for future expansion. Shell space is enclosed by the building envelope but otherwise left unfinished.

Short-Term Bicycle Storage Nonenclosed bicycle parking typically used by visitors for a period of 2 hours or less.

Simple Box Energy Modeling Analysis Also known as "building-massing model energy analysis," this is a simple base-case energy analysis that informs the team about the building's likely distribution of energy consumption and is used to evaluate potential project energy strategies. A simple box analysis uses a basic, schematic building form.

Site Assessment An evaluation of an area's above-ground and subsurface characteristics, including its structures, geology, and hydrology. Site assessments typically help determine whether contamination has occurred and the extent and concentration of any release of pollutants. Remediation decisions rely on information generated during site assessments.

Site Master Plan An overall design or development concept for the project and associated (or potentially associated) buildings and sites. The plan considers future sustainable use, expansion, and contraction. The site master plan is typically illustrated, with building plans (if applicable), site drawings of planned phased development, and narrative descriptions.

Softscape The elements of a landscape that consist of live, horticultural elements.

Soft Space An area whose functions can be easily changed. For example, hospital administrative offices could be moved so that this soft space could be converted to a laboratory. In contrast, a lab with specialized equipment and infrastructure would be difficult to relocate.

Solar Garden/Community Renewable Energy System Shared solar array or other renewable energy system with grid-connected subscribers who receive credit for the use of renewables using virtual net metering. (Adapted from solargardens.org.)

Solar Reflectance (SR) The fraction of solar energy that is reflected by a surface on a scale of 0 to 1. Black paint has a solar reflectance of 0; white paint (titanium dioxide) has a solar reflectance of 1. The standard technique for its determination uses spectrophotometric measurements, with an integrating sphere to determine the reflectance at each wavelength. Determine the SR of a material by using the Cool Roof Rating Council Standard (CRRC-1). The ratio between the solar energy globally reflected by a surface and the total incident solar energy. Traditionally, roofing material's SR can be between 0.05 and 0.5 (e.g., 0.05 for black membrane roofs, 0.08–0.5 for metal roofs, and 0.2 for clay tiles).

Solar Reflectance Index (SRI) A measure of the constructed surface's ability to stay cool in the sun by reflecting solar radiation and emitting thermal radiation. It is defined such that a standard black surface (initial solar reflectance 0.05, initial thermal emittance 0.90) has an initial SRI of 0, and a standard white surface (initial solar reflectance 0.80, initial thermal emittance 0.90) has an initial SRI of 100. To calculate the SRI for a given material, obtain its solar reflectance and thermal emittance via the Cool Roof Rating Council Standard (CRRC-1). SRI is calculated according to ASTM E 1980. Calculation of the aged SRI is based on the aged tested values of solar reflectance and thermal emittance.

Sound-Level Coverage A set of uniformity criteria that ensure consistent intelligibility and directionality of audible frequencies for all occupants within a space.

Source Reduction A decrease in the amount of unnecessary material brought into a building in order to produce less waste. For example, purchasing products with less packaging is a source reduction strategy.

Spatial Daylight Autonomy (sDA) A metric describing annual sufficiency of ambient daylight levels in interior environments. It is defined as the percentage of an analysis area (the area where calculations are performed, typically across an entire space) that meets a minimum daylight illuminance level for a specified fraction of the operating hours per year (i.e., the Daylight Autonomy value following Reinhart & Walkenhorst, 2001). The illuminance level and time fraction are included as subscripts, as in $sDA_{300/50\%}$. The sDA value is expressed as a percentage of area. In the case of $sDA_{300/50\%}$, it indicates that a certain percent of area must meet or exceed 300 lux for at least 50% of the working hours per year.

Spatial Daylight Autonomy ($sDA_{300/50\%}$) The percentage of analysis points across the analysis area that meet or exceed this 300-lux value for at least 50 percent of the analysis period.

Speech Privacy A condition in which speech is unintelligible to casual listeners (ANSI T1.523–2001).

Speech Spectra The distribution of acoustic energy as a function of frequency for human speech.

Streetcar A transit service with small, individual rail cars. Spacing between stations is uniformly short and ranges from every block to ¼ mile, and operating speeds are primarily 10–30 miles per hour (15–50 kilometers per hour). Streetcar routes typically extend 2–5 miles (3–8 kilometers).

Structure Elements carrying either vertical or horizontal loads (e.g., walls, roofs, and floors) that are considered structurally sound and nonhazardous.

Systems Manual Provides the information needed to understand, operate, and maintain the systems and assemblies within a building. It expands the scope of the traditional operating and maintenance documentation and is compiled of multiple documents developed during the commissioning process, such as the owner's project requirements, operation and maintenance manuals, and sequences of operation.

T

Technical Release (TR) 55 An approach to hydrology in which watersheds are modeled to calculate storm runoff volume, peak rate of discharge, hydrographs, and storage volumes, developed by the former U.S. Department of Agriculture Soil Conservation Service.

Thermal Emittance The ratio of the radiant heat flux emitted by a specimen to that emitted by a black-body radiator at the same temperature. (Adapted from the Cool Roof Rating Council.)

Three-Year Aged Solar Reflectance (SR) A solar reflectance or solar reflectance index rating that is measured after 3 years of weather exposure. The SR is also known as the Solar Reflectance Index (SRI) value.

Time-of-Use Pricing An arrangement in which customers pay higher fees to use utilities during peak time periods and lower fees during off-peak time periods.

U

Undercover Parking   Vehicle storage that is underground, under deck, under roof, or under a building.

Uninterruptible Power Supply (UPS) Output   The electricity provided by a unit that keeps information technology (IT) equipment functioning during a power outage. UPS output does not include efficiency losses from the unit itself but does include losses from downstream electrical distribution components, such as power distribution units, and it may include non-IT ancillary devices installed in IT racks, such as fans. If the UPS system supports non-IT equipment (e.g., computer room air-conditioning units, computer room air handlers, in-row coolers), this usage must be metered and subtracted from the UPS output reading. The metering approach should be consistent with the metering required for the power usage efficiency (PUE) category (e.g., continuous consumption metering for PUE Categories 1, 2, and 3).

Universal Waste   Hazardous items that are easily purchased and commonly used. Examples include batteries, pesticides, mercury-containing equipment, and light bulbs. See epa.gov/osw/hazard/wastetypes/universal/index.htm.

Unoccupied Space   An area designed for equipment, machinery, or storage rather than for human activities. An equipment area is considered unoccupied only if retrieval of equipment is occasional.

Upstream Equipment   A heating or cooling system or control associated with the district energy system (DES) but not part of the thermal connection or interface with the DES. Upstream equipment includes the thermal energy conversion plant and all the transmission and distribution equipment associated with transporting the thermal energy to the project building or site.

USDA Organic   The U.S. Department of Agriculture's (USDA) certification for products that contain at least 95 percent ingredients (excluding water and salt) produced without synthetic chemicals, antibiotics, or hormones. Any remaining ingredients must consist of USDA-approved nonagricultural substances or agricultural products that are not commercially available in organic form.

V

Vertical Illuminance   Illuminance levels calculated at a point on a vertical surface or that occur on a vertical plane.

Vision Glazing   The glass portion of an exterior window that permits views to the exterior or interior. Vision glazing must allow a clear image of the exterior and must not be obstructed by frits, fibers, patterned glazing, or added tints that distort color balance.

W

Walk-Off Mats   Mats placed inside the building entrances to address pollution point source control by capturing dirt, water, and other materials tracked inside by people and equipment.

Walking Distance   The distance that a pedestrian must travel between origins and destinations without obstruction, in a safe and comfortable environment on a continuous network of sidewalks, all-weather-surface footpaths, crosswalks, or equivalent pedestrian facilities. The walking distance must be drawn from an entrance that is accessible to all building users.

Waste Diversion   A management activity that disposes of waste through methods other than incineration or landfilling. Examples include reuse and recycling.

Waste to Energy   The conversion of nonrecyclable waste materials into usable heat, electricity, or fuel through a variety of processes, including combustion, gasification, pyrolization, anaerobic digestion, and landfill gas (LFG) recovery.

Wastewater   Water that has been used for a purpose and conveyed by building plumbing systems toward a point of treatment and disposal. Wastewater from buildings can be classified as gray water, black water, or process wastewater.

Water Body   The surface water of a stream (first-order and higher, including intermittent streams), arroyo, river, canal, lake, estuary, bay, or ocean. It does not include irrigation ditches.

Water Budget   A project-specific method of calculating the amount of water required by the building and associated grounds. The budget takes into account indoor, outdoor, process, and makeup water demands and any on-site supply, including estimated rainfall. Water budgets must be associated with a specified amount of time, such as a week, month, or year, and a quantity of water, such as kilogallons or liters.

Wet Meter   A device installed inside a water pipe to record the volume of passing water.

Wetland   An area that is inundated or saturated by surface or groundwater at a frequency and duration sufficient to support, and that under normal circumstances does support, a prevalence of vegetation typically adapted for life in saturated soil conditions. Wetlands generally include swamps, marshes, bogs, and similar areas, but exclude irrigation ditches unless delineated as part of an adjacent wetland.

Wood   Plant-based materials that are eligible for certification under the Forest Stewardship Council. Examples include bamboo and palm (monocots) as well as hardwoods (angiosperms) and softwoods (gymnosperms).

X

Xeriscaping   Landscaping that does not require routine irrigation, which uses drought-adaptable and low-water plants as well as soil amendments such as compost, mulch, and rocks to reduce evaporation and make irrigation unnecessary.

Y

Yard Tractor   A vehicle used primarily to facilitate the movement of truck trailers and other types of large shipping containers from one area of a site to another. It does not include forklift trucks. Also known as terminal tractor, yard truck, utility tractor rig, yard goat, or yard hustler.

Z

Zero-Emission Vehicle (ZEV)   A vehicle that does not emit exhaust gas or other pollutants from the onboard source of power.

Zero Lot Line Project   A plot whose building footprint typically aligns or nearly aligns with the site limits.

Mock-up Exam

1. Which impact category emphasizes helping people with the aspect of the triple bottom line?

   a. Promote sustainable and regenerative material resources cycles

   b. Protect, enhance, and restore biodiversity and ecosystem services

   c. Enhance social equity, environmental justice, and community quality of life

   d. Reverse contribution to global climate change

2. What does the integrative process in a project encourage?

   a. Life-cycle assessments

   b. Enhanced social equity, environmental justice, community health, and equality of life

   c. Team collaboration

   d. Regenerative systems promotion

3. What is the premise of combining people, planet, and profit?

   a. LEED professionals

   b. Stakeholders

   c. Shareholders

   d. The environment

4. The Discovery phase occurs _____ the schematic design.

   a. Before

   b. After

   c. During

   d. Simultaneously with

5. Which certificate does a project that earned 58 points receive?

   a. Certified

   b. Silver

   c. Gold

   d. Platinum

6. A project team is trying to decide on selecting the appropriate rating system (LEED BD+C or LEED O&M) for a retail store. What should the team do?

   a. Choose LEED BD+C

   b. Choose LEED O&M

   c. Choose both to earn extra credit

   d. Use the 40/60 rule

7. Select all the incorrect statements from the following list. Choose three.

   a. Prerequisites are optional.

   b. Two points are awarded for each prerequisite.

   c. Prerequisites are mandatory.

   d. No points are awarded for prerequisites.

   e. Prerequisites are not required if credits in that particular category are all met.

8. Which of the following is a certification and credentialing body within the green business and sustainability industry that administers the project certifications and professional credentials?

   a. LEED

   b. LEED with specialty

   c. USGBC

   d. GBCI

9. The Multifamily Midrise rating system is used for which category?

   a. Neighborhood Development

   b. Building Operation and Maintenance

   c. Homes

   d. Interior Design and Construction

10. Which rating system was developed specifically for the tenant fit out?

    a. LEED BD+C

    b. LEED ID+C

    c. LEED ND

    d. LEED Homes

11. Select the LEED program that simplifies the certification process for a large number of new and existing construction projects that do not have to have the same floor plan but do have to belong to a single organization?

    a. Recertification Program

    b. LEED Professional Credentials

    c. LEED Campus Program

    d. LEED Volume Program

12. What decides the level of certification on a project?

 a. The type of the rating system

 b. The number of LEED-accredited people on the team

 c. The number of credit points earned

 d. The size and budget of the project

13. How many total credit points can Location and Transportation contribute toward achieving LEED certification?

 a. 10

 b. 12

 c. 16

 d. 13

14. Which one of the following is NOT the intent of the site location?

 a. Greenhouse gas emissions

 b. The incidence of obesity

 c. Heart disease

 d. Hypertension

 e. Fuel consumption

15. Infill sites are required in historical areas. Select an example from the following.

 a. A high-rise building in a previously developed site

 b. Brownfield site

 c. Right of way

 d. Forest

16. List three sensitive sites mentioned in the LT category?

 a. Brownfield

 b. Prime farmland

 c. Wetland

 d. Floodplains

17. When determining the total parking capacity for a project, which of the following is not included?

 a. New and existing surface parking spaces

 b. New and existing garage or multilevel parking spaces

 c. Any off-street spaces outside the project boundary that are available to the building's users

 d. Parking spaces for fleet and inventory vehicles unless these vehicles are regularly used by employees for commuting as well as business spaces

18. What is the percentage of the preferred parking for the total parking spaces?

    a. 2 percent

    b. 5 percent

    c. 3 percent

    d. 10 percent

19. What is a greenfield?

    a. Land that has been previously developed

    b. A gas station

    c. Land that is covered with grass and underneath it is a parking lot

    d. Land that has been previously undeveloped

20. Which of the following are pedestrian amenities? Select three.

    a. Street trees

    b. Shades

    c. Water fountains

    d. Highways and bridges

21. The LT category is undertaken during which phase of the project?

    a. Construction

    b. Postconstruction

    c. Design

22. In the LT category, promoting connectivity is one of the essential movements a LEED professional must consider. What are some of the community connectivity promotions?

    a. Limiting cul-de-sacs

    b. Prohibiting gated communities

    c. Using street grid patterns

    d. Opening up new transportation routes with overlaps

23. What is development density in the LT category measured by?

    a. Floor-area ratio (FAR)

    b. Square feet per acre

    c. By footprint

    d. All of the above

24. Building a project in a remote area is not highly encouraged by the LEED community. How should the environmental impact be mitigated?

    a. Develop more projects in this remote area

    b. Promote carpooling

    c. Build sidewalks with previous payment

    d. None of the above

25. The purpose of density calculation is to ___ .

    a. Determine how many people will be carpooling to work versus riding in their own vehicles

    b. Determine transportation routes to and from a certain location at peak travel times

    c. Determine the square footage of buildings per acre

    d. Calculate the density of the topsoil during a brownfield remediation.

26. The Sustainable Sites category addresses all except which?

    a. Rainwater runoff

    b. Heat island effect

    c. Light pollution

    d. Energy and atmosphere

27. The SS Credit: Open Space requires outdoor space greater than or equal to what percentage?

    a. 20 percent

    b. 40 percent

    c. 30 percent

    d. 32 percent

28. Choose three strategies to reduce rainwater runoff.

    a. Using porous pavement

    b. Including a rain garden

    c. Collecting rainwater

    d. Designing for a larger footprint

29. A building has a 65 percent vegetated roof. The project team decided to replace 30 percent of the vegetated roof with solar reflectance index (SRI). Which credit would this change help with?

    a. Construction activity pollution prevention

    b. Renewable energy

    c. Heat island reduction

    d. Sensitive land

30. In the SS prerequisite, Schools and Healthcare rating systems must perform an assessment, which includes Phase l and Phase II per (ASTM E1527). What type of assessment is that?

    a. Site assessment

    b. Environmental site assessment

    c. Biodiversity site assessment

    d. Community connectivity assessment

    e. Neighborhood development assessment

31. LEED professionals use the concept "The area on a project site used by a building footprint can be maximized without sacrificing square footage by building 'up rather than out'" to describe which of the following?

    a. Green infrastructure (GI)

    b. FAR

    c. Low-impact development (LID)

    d. SRI

32. How can we reduce light pollution?

    a. Install more storefront

    b. Install motion sensors and timers

    c. Install louvers at the exterior windows

    d. Install interior curtains

33. An architect wants to replace a portion of a fully vegetated roof with solar panels. Which of the credits below will be enhanced?

    a. Rainwater management

    b. Heat island effect

    c. Lighting power density

    d. Enhanced refrigerant management

34. A project team is reviewing credit options to synergize two credits to maximize reward points. Which of the following can be synergized?

    a. Demand response and high-priority site

    b. Sensitive land protection and low-emitting materials

    c. Green vehicles and thermal comfort

    d. Daylight and quality views

35. Typically, newly built buildings require adding piles of soils to the site. What environmental benefit would this practice provide?

    a. Maximizing open space

    b. Restoring habitat

    c. Increasing site density

    d. Preventing construction pollution

36. Who designs the erosion and sedimentation control plan for projects?

    a. Project manager

    b. Owner

    c. LEED professional

    d. Licensed engineer

37. Rainwater runoff causes which natural environmental issue?

    a. Erosion

    b. Sick building syndrome

c. Inside contamination

d. Carbon footprint

38. What measure is used for water usage for flow fixtures?

a. Per flush

b. Per minute

c. Per second

d. None of the above

39. The ultra-low-flow urinal fixture type uses an optimized valve that produces a pressure-assisted flush. What is the approximate water usage for the valve?

a. 1.28 gpf

b. 1.28 gpm

c. 0.125 gpf

d. 0.125 gpm

40. What is the purpose of the design case in the WE category?

a. The design will achieve reductions in water use from the baseline water use calculation.

b. The design creates a list of project team participants.

c. The design will achieve an increase in water use from the baseline water use calculation.

41. The main goal of WE Prerequisite: Indoor Water Use Reduction and WE Credit: Indoor Water Use Reduction is to reduce water consumption by what percentage?

a. 20 percent

b. 30 percent

c. 40 percent

d. 50 percent

42. What is the default gender ratio for full-time equivalent (FET) occupants?

a. 40:60

b. 70:30

c. 50:50

d. 90:10

43. A building owner is stressing the reduction of outdoor water use, so the project team decided to install which of the following?

a. Drip irrigation

b. Mulching

c. Submetering

d. Xeriscaping

44. Which is a flush fixture?

    a. Dishwasher

    b. Showerhead

    c. Urinals

    d. Kitchen sink

45. What is wastewater that has been treated and purified from no potable use?

    a. Black water

    b. Gray water

    c. Reclaimed water

    d. Cooling tower blowout

46. Which calculates the number of occupants in a building with associated average daily water fixture use?

    a. Water closet

    b. EPAct 1992

    c. FTE

    d. Drip irrigation

47. How many points are available in the WE category?

    a. 10

    b. 16

    c. 11

    d. 13

48. What is the water measurement unit of faucets and showers?

    a. Gallons per flush (gpf) or liter per flush (Lpf)

    b. Gallons per minute (gpf) or liter per minute (Lpf)

    c. Gallons per hour (gph) or liter per hour (Lph)

    d. Gallons per flush gpf or liter per flush (Lpf)

49. Which woud be an example of black water? Select two.

    a. Wastewater from urinals

    b. Collected rainwater

    c. Water from dishwasher

    d. Water from the shower

50. What is full-time equivalent (FTE) an example of?

    a. A full-time front desk worker

    b. A facility visitor

    c. A maintenance person on call who works for 2 hours

    d. An area supervisor who works remotely

51. Identify an approach that helps to reduce indoor and outdoor water use.

    a. A green roof

    b. Parking space partially underground

    c. Main metering and submetering

    d. Collection of rainwater

52. Which is NOT a passive energy design?

    a. Solar energy

    b. Daylight

    c. Natural ventilation

    d. Mechanical ventilation

53. Renewable energies include all except which of the following?

    a. Solar power

    b. Wind power

    c. Geothermal bioenergy

    d. Waste-to-energy systems

    e. Passive solar architectural features

54. The intent of refrigerant management is to eliminate the use of which of the following?

    a. HCFC

    b. HFC

    c. CFC

    d. CO_2

55. A commissioning authority (CxA) must do all the following except?

    a. Review the OPR, DOB, and project design

    b. Develop and implement a Cx plan

    c. Develop construction checklists

    d. Design the plumbing system for rainwater harvesting

56. As part of the refrigerant elimination management, CFC use is banned and HCFC is phased out by which of the following?

    a. Illuminating Engineering Society of North America (IESNA)

    b. Montreal Protocol

    c. Commissioning authority (CxA)

    d. Energy Star portfolio manager

57. Select two projects ineligible for LEED certification.

    a. A new project that has no mechanical system and relies 100 percent on natural ventilation

    b. An existing building with a phase-out plan for CFC-based refrigerants

    c. A new building that uses CFC-based refrigerants

    d. An existing building with no phase-out plan for CFC

58. What helps utility companies optimize their supply-side energy and delivery systems?

    a. Renewable energy

    b. REC

    c. Demand response

    d. Energy use intensity (EUI)

59. According to the Montreal Protocol, which is the least desired refrigerant system and has been banned?

    a. CFC

    b. HFC

    c. HCFC

    d. HCFC-22

60. An architect is designing a building to reduce energy demand, so the architect must follow the LEED consultant's advice. What is the advice? Select two.

    a. Incorporate passive design

    b. Use high-efficiency glazing to retain heat

    c. Convince the owner to reduce the size of the building

    d. Purchase land that is near an energy source

61. How many points are available in the Energy and Atmosphere category?

    a. 16

    b. 11

    c. 33

    d. 9

62. A building's HVAC system uses chlorofluorocarbon, so the owner wants to conduct major renovation of the building and apply for the LEED new construction certification. What is true about this process?

    a. The building will not be granted any LEED certification.

    b. The building will be granted only the LEED Certificate level.

    c. The owner must implement a phase-out plan to earn LEED certification.

    d. The owner must buy a new HVAC system with the same refrigerant.

63. If a company is limited due to on-site installation costs and the ability of their utility provider to sell them green power, the company can support the production of renewable energy and offset their carbon footprint by doing what?

    a. Enrolling in the Green-e certification program

    b. Purchasing a renewable energy certificate

    c. Appealing to the court by requesting a variance

    d. Specifying high-efficiency appliances

64. Which program demands building owners decrease electricity use?

    a. Demand response

    b. Energy Star portfolio manager

    c. Montreal Protocol

    d. Net-zero energy

65. What is the mutual factor between REC and carbon offsets?

    a. Both of them can be sold

    b. Both of them can be purchased

    c. Both of them can be traded off

    d. They do not have a mutual factor

66. Which is defined by a set of specific rules, requirements, and guidelines based on ISO specifications for developing environmental declarations for one or more product categories?

    a. Environmental Product Declaration (EPD)

    b. Product Category Rule (PCR)

    c. Health Product Declarations (HPD)

    d. Whole-building LCA

67. Which approach of the MR Credit: Building Life Cycle Impact Reduction is used for new buildings?

    a. Historic building reuse

    b. Renovation of abandoned or blighted building

    c. Building and materials reuse

    d. Whole-building LCA

68. Low-emitting materials for the interior and exterior of buildings are organized into seven categories for VOCs, each with different thresholds of compliance. Which one does NOT belong?

    a. Interior paints and coating, adhesives, and sealants

    b. Flooring, composite wood, furniture

    c. Ceiling, walls, and acoustic insulation

    d. Exterior applied products

    e. Piping and pipe insulation

69. Life-cycle approach and materials waste management are two of the four goals of which category?

    a. LT

    b. SS

  c. MR

  d. WE

70. How many points are available in the MR category?

  a. 11

  b. 33

  c. 16

  d. 13

71. EPD products must be aligned and in compliance with which of the following?

  a. ASHRAE 55-2010

  b. ASHRAE 90

  c. International Organization for Standardization (ISO)

  d. City or county code

72. Select two types from below that are preconsumer recycled content.

  a. Chemical plant hazardous waste

  b. Office paper

  c. Sawdust

  d. Construction debris

73. Select one classification from the following that fits the construction debris under the Materials and Resources Category?

  a. Preconsumer recycled content

  b. Construction debris should never be used

  c. Construction materials

  d. Postconsumer recycled content

74. What is the aim of the construction waste management plan?

  a. To divert materials from waste stream

  b. To increase the number of trash bins in buildings

  c. To strategically locate trash bins around accessible locations

  d. To implement recycling plans more often

75. A project team has hired a LEED professional for consultation on reducing construction waste and the need for new building materials when renovating an office building between tenants moving in or out. Which option is more suitable?

  a. Hiring skilled trades

  b. Using only prefabricated products for interior and exterior

  c. Designing for flexibility of configuration

  d. Using a commingled materials stream

76. A LEED professional is working with his or her team on completing project documents during the design phase. What is one way to enforce using the specifically selected materials?

    a. Calling the subcontractors who are going to build the project and communicating the information to them.

    b. Communicating the specified material during the 4-hour charrette.

    c. Including the materials' details in the specifications.

    d. It is the responsibility of the subcontractor to figure it out and not the reasonability of the LEED professional.

77. Which of the following certificates has five levels: basic, bronze, silver, gold, and platinum?

    a. Cradle-to-cradle certificate

    b. Cradle-to-gate certificate

    c. Cradle-to-grave certificate

    d. Environmental product declaration

78. City officials have signed a new act to offer incentives for owners who reduce sending waste to landfills. How will the owner qualify for these incentives?

    a. Use materials with low life-cycle costs

    b. Recycle materials

    c. Use materials within 50 miles of the project

    d. Eliminate sick building syndrome

79. A business owner who owns a barn recovered its wood and used it in a new office project that was recently developed. Which type of wood is this considered?

    a. New wood obtained from a demolition project

    b. Recycled and refurbished materials

    c. Reused wood

    d. Wood containing recycled materials

80. What is the measure for ventilation (or airflow rate)?

    a. Minute per gallon (mpg)

    b. Megawatt-hour (MWh)

    c. Cubic feet per minute (cfm)

    d. Energy use intensity (EUI)

81. LEED projects use _____ as the standard to measure thermal comfort.

    a. ASHRAE 62.1-2013

    b. ASHRAE 55-2010

    c. SMACNA

    d. VOC

82. What is it called when people have symptoms such as headaches, throat irritation, dizziness, and nausea due to poor ventilation with unfiltered air?

    a. Health product declarations

    b. Sick building syndrome

    c. Chain of custody

    d. Registration, evaluation, authorization, and restriction of chemicals (REACH)

83. Which is a manual that defines the acceptable process for the natural ventilation of nondomestic buildings?

    a. Circadian rhythms

    b. CIBSE AM10

    c. ANSI/BIFMA e3-2011

    d. Flush-out

84. Smoking should only be allowed outside the building in designated smoking areas located at least ___ feet from all entries.

    a. 15

    b. 20

    c. 25

    d. 50

85. EQ Credit: Quality Views aims to give building occupants a connection to the natural outdoor environment by providing quality views. To achieve this credit, a direct line of sight to the outdoor via vision glazing must be at least ___ .

    a. 50 percent

    b. 75 percent

    c. 25 percent

    d. 30 percent

86. All of the following are hard costs except which two?

    a. Permitting fee

    b. Steel structure

    c. Direct cost

    d. Rental equipment fee

    e. Consulting fee

87. Reducing the lighting power density for a project can reduce energy use, which is the result of which of the following?

    a. Daylighting

    b. Use of minimum efficiency reporting value

    c. Mechanical ventilation

    d. Volatile organic compound

88. A delivery of batt insulation is made by the supplier. Because the delivery took place during a rainy day, some of the materials got wet. What should the contractor do?

    a. Wait until the materials dry out, then install them.

    b. Install the materials and then use fans to dry them out.

    c. Dispose of them because the materials became useless once exposed to moisture.

    d. Ask for a replacement.

89. Postconstruction, during building operation, select an option that helps improve indoor air quality.

    a. Flush out the building with 10,000 cfm after occupants move in

    b. Use a highly volatile organic compound

    c. Establish a green cleaning policy

    d. Provide thermal comfort

90. One of the required categories in the thermal comfort control is to provide ___ .

    a. Individual thermal comfort controls for at least 90 percent of individual occupant spaces and to group thermal comfort controls for all shared multioccupant spaces

    b. Individual thermal comfort controls for at least 50 percent of both individual occupant spaces and shared multioccupant spaces

    c. Individual thermal comfort controls for at least 50 percent for shared multioccupant spaces

    d. Individual thermal comfort controls for at least 50 percent of individual occupant spaces and group thermal comfort controls for all shared multi-occupant spaces

91. A project team had a conflict of opinion on designing an open floor concept. The project team decided to install partitioned cubes with frosted wall panels. After consulting with a LEED professional, the project team found out their choice of the partitioned cubes with frosted wall panels would negatively affect which of the following?

    a. Occupancy sensors

    b. Acoustic performance

    c. Quality views

    d. Air movement and circulation

92. Select two primary factors for thermal comfort according to ASHRAE 55-2010.

    a. Surface temperature

    b. Acoustics

    c. Light reflection

    d. Clothing

93. During the design phase, the team is making decisions on providing views to the outside to most employees and also using as much daylight as possible. What design strategy should they follow?

    a. Locating custodial and storage areas on the perimeters and cubical offices in the center with low wall partitions

    b. Locating custodial and storage areas on the perimeters and cubical offices in the center without wall partitions

    c. Placing conference rooms in the center with open floor concept offices

    d. Maximizing the curtain wall ratio to the floor regardless of the floor plan design

94. Synergy is a common practice in LEED projects to earn more points. Which could be an example?

    a. Quality transit and quality view access

    b. Daylight and interior lighting

    c. Water metering and demand response

    d. Bicycle facilities and indoor air quality assessment

95. Which application helps acoustic performance?

    a. Installing double-pane windows

    b. Installing shading devices and interior curtain

    c. Installing acoustic tiles

    d. Using a MERV-13 filter system

96. A project can earn points under EQ Credit: Indoor Air Quality Assessment if air quality testing takes place ___ .

    a. Before occupancy

    b. After occupancy

    c. Before construction commences

    d. During occupancy

97. Which category focuses on the most important environmental issues affecting a particular locality?

    a. Innovation

    b. Water Efficiency

    c. Regional Priority

    d. Indoor Environmental Quality

98. What is the intent of the IN category?

    a. To earn extra points only

    b. To encourage projects to achieve exceptional and/or innovative performance

    c. To fulfill collaboration and teamwork

    d. None of the above

99. What is the intent of the Regional Priority credit?

    a. To provide an incentive for the achievement of credits that address geographically specific environmental, social equity, and public health priorities

    b. To reduce the amount of raw materials used in construction

    c. To accomplish the same amount of work with less energy expended

    d. To reduce the footprint

100. A maximum of 4 points can earn for which of the choices that follow?

    a. Innovation

    b. Water Efficiency/Indoor Water Use Reduction

    c. Regional Priority

    d. Innovation/Innovation

Mock-up Exam: Answers

1. **c**

2. **c**

3. **b** The focus of the triple bottom line is the stakeholders.

4. **a**

5. **b** Silver certification is between 50 and 59 reward points.

6. **d** When a project team is unsure of which rating system should be selected, it is recommended to use the 40/60 rule.

7. **a, b, e**

8. **d** GBCI is a third-party organization that provides independent oversight of professional credentialing and the project certification program related to green building.

9. **c**

10. **b**

11. **d** The LEED volume program certifies at least 25 projects or more that do not have the same design and do not have to be owned by the same organization.

12. **c** The number of points earned is determined by the level of certification.

13. **c**

14. **e** All options are correct as LT category's intent except option e (fuel consumption). Some may argue fuel consumption is necessary to be addressed, and I agree, but this question is about the bigger picture. Burning fossil fuel and the process that takes place have a significant impact on the environment, but not fuel consumption.

15. **a** Infill location is defined as a real estate development site that exists within a mostly built-out market. Usually located within an urban area, infill locations look to fill the few vacant lots that exist between other developments in the area.

16. **b, c, d**

17. **d**

18. **b**

19. **d**

20. **a, b, c**

21. **c**

22. **a, b, c** Community connectivity is achieved by removing physical barriers that might hamper interpersonal communication and the ability to walk or bicycle. Options a, b, and c help accomplish the goal; however, option d should not be used because opening new routes with overlap is not encouraged.

23. **b** Development density is the total square footage of all buildings within a particular area.

24. **b** Carpooling is one of the strategies to limit the impact of transportation on the environment.

25. **c**

26. **d**

27. **c**

28. **a, b, c** Option d increases water runoff. The larger the footprint is, the more runoff there will be.

29. **c** This credit encourages the project team to design a building and site that have minimal contributions to heat island effects.

30. **b**

31. **b** FAR is the concept of building up rather than out and maximizes the building footprint without cutting square feet.

32. **a & b** Smart light design can reduce light trespass significantly and includes options a and b.

33. **b** Option b is not a credit. Options a and d are irrelevant credits.

34. **d** Option d is the correct answer, so the team can synergize daylight with quality views by maximizing window size to the floor plan and also by orienting the building strategically.

35. **b** Open space intends to create exterior open space that encourages interaction with the environment, social interaction, passive recreation, and physical activities. Increasing site density is to build up, not out. The prevention of construction pollution is to control soil erosion, waterway sedimentation, and generation of airborne particulates. Therefore, option b, which is restoring habitat, is the best option because restoring habitat is restoring soil to support biodiversity.

36. **d** A licensed civil engineer is responsible for designing the sedimentation control plan. The owner approves it for financial reasons, and the project manager implements the design; the LEED professional verifies the requirements for accreditation.

37. **a**

38. **b**

39. **c**

40. **a**

41. **a**

42. **c**

43. **d**

44. **c**

45. **c**

46. **c**

47. **c**

48. **b**

49. **a & c**

50. **a**   FTE is used to calculate the number of occupants in a building with associated average daily water fixture use.

51. **d**   Rainwater collection is the only approach among these options that helps reduce potable water use for both indoors and outdoors.

52. **d**   The definition of passive energy design is that windows, walls, and floors are made to collect, store, reflect, and distribute solar energy in the form of heat in the winter and to reject solar heat in the summer.

53. **e**

54. **c**

55. **d**

56. **b**

57. **c & d**

58. **c**   The demand response (DR) program helps project teams think more holistically about their building and consider the interconnection between building systems. DR is a program offered by participating utilities nationwide to encourage large energy users to reduce energy loads during peak energy usage times in exchange for reduced rates.

59. **a**

60. **a & b**

61. **c**

62. **c**   Option d, buying a new HVAC system, does not change the fact the refrigerant (CFC) must be banned. As long as the building uses CFC refrigerant, it will not be granted any LEED certification, so the only solution is to phase out the CFC.

63. **b**   A renewable energy certificate is a form of green power and can be purchased to offset carbon.

64. **a**   According to the USGBC, demand response allows utilities to call on buildings to decrease their electricity use during peak times, reducing the strain on the grid and the need to operate more power plants, thus potentially avoiding the cost of constructing new plants.

65. **b**

66. **b**

67. **d**   All the options listed are ways used for the MR Credit building life cycle; however, option d is the only approach utilized for "new buildings."

68. **e**   Check out the low-emitting materials for VOC product categories in the IEQ chapter.

69. **c**

70. **d**

71. **c**   Environmental Product Declaration (EPD) are building materials criteria and must conform to a specific standard of the International Organization for Standardization (ISO) and have at least a cradle-to-gate scope that are valued at one-half to a whole product, depending on the type. Therefore, option c is the correct answer. Option d is incorrect because codes in cities and counties vary, which defeats the purpose of having a standardized code under ISO. Options a and b are irrelevant. Option a is for thermal environment and option b is about energy efficiency.

72. **b & c**   Options b and c are preconsumer recycled contents; they are made from manufacturers' waste that never actually made it to the consumer. Sawdust can be made into walls, floors, and wood products. Office paper can be made into paper towels, napkins, greeting cards, or cardboard. Option d is postconsumer, and option a should be reused.

73. **d**   Check out the definition of postconsumer recycled content.

74. **a**   The USGBC made construction waste management an important first step of every LEED rating system by requiring a construction waste plan for every project. This is represented by waste diversion.

75. **c**   This question is tricky, and it is meant to mislead you. Pay attention to the phrase "moving in or out" and to "more suitable." More suitable means there should be other options applicable, but choose the best option that has a long-term positive effect. Options a, b, and c are all good and could be used, but option c has a significant effect on reducing material waste throughout the life span of the project. Moving in or out is for design flexibility and easy configuration to allow quick renovation and a module approach.

76. **c**   The c option is the correct one. All other options might be correct also, but for another scenario.

77. **a**

78. **b**   Using materials with lower life-cycle costs helps reduce operating costs and maintenance over the product's lifetime, so it is not applicable. Using materials with a within 50 miles of the project helps lower transportation costs of moving materials, so it is more for economic reasons. Option d is not applicable, so option b is the best answer. Recycling materials reduces materials sent to landfills.

79. **c**   Option b is wrong because the wood flooring is not "recycled." Option c is the correct answer. Option d is obviously wrong because the wood does not contain recycled materials.

80. **c** The unit measure of airflow is cfm.

81. **b** Thermal comfort follows the ASHRAE 55-2010 standard.

82. **b**

83. **b** CIBSE AM10 is a manual standard and contains a flow diagram, which helps to design natural ventilation.

84. **c**

85. **b**

86. **a & e**

87. **a** Option b is the MERV filter standard. Option c increases the energy and is not relevant to lighting. Option d has to deal with the MR category.

88. **d** This is a judgment call by the contractor. To ensure the materials did not get moldy and so the owner is without responsibility, it is better to return the materials.

89. **c** The flush out must be done before or during occupancy, not after occupancy (option a). Low-VOC materials are recommended, not high VOC. Thermal comfort has to deal more with temperature than air quality. Option c is the most ideal because it eliminates material-based chemicals. The best strategy to solve questions like this is by choice elimination.

90. **d** The question is meant so close attention is paid to percentage of individual spaces, which is 50% vs. shared spaces, which is the entire space.

91. **c** Partitioned cubes with frosted wall panels (especially high walls) affect view quality. The other options do not interfere with wall panels.

92. **a & d** There are six thermal factors (according to SHRAE 55-2010): clothing, air movement, metabolic rate, humidity, air temperature, and surface temperature.

93. **c** The synergy approach is between daylight and quality view.

94. **b** There is a synergy between daylight and interior light. When more daylight is let in, fewer artificial lights are needed, thus saving energy. There are no synergies between the other options.

95. **c** Acoustic tiles enhance acoustic performance.

96. **a** Points are earned for this credit (EQ Credit: Indoor Air Quality Assessment) by either conducting a flush out of the building before or during occupancy or by testing the air quality before occupancy to ensure the air is safe to breathe.

97. **c** Regional Priority given to provide an incentive for the achievement of credits that address a geographically specific environment.

98. **b**

99. **a**

100. **c** Check out the scorecard.

Real-World Case Studies

Involving Stakeholders at Early Stage of the Design Process to Improve Credit Points Allocation

Husam A. Alshareef, Krystal Vallejos, Patrick Aguilar,
Dallas Leasure, Chance Trueblood

Abstract

Several sustainable buildings (existing or new construction) that seek Leadership in Energy and Environmental Design (LEED) credentials tend to lack proactive plans at a very early stage of the project. Stakeholders and decision-makers typically wait until the design phase begins to discuss LEED's categories (i.e. credit point allocations), which drains out the budget with a limited number of resources. This conventional method has a higher probability of reducing production and collaboration and also limits creativity and innovation. Therefore, this research is intended to evaluate the early preparation of eco-friendly buildings (e.g. the Technology Building at Colorado State University (CSU-P) as a case study) and examine practical applications to seek a LEED certification.

A collaborative iterative process approach was utilized by researching and evaluating ideas and conducting interviews with stakeholders and decisionmakers. This process is undertaken to identify the most useful materials, items, ideas, and then weigh them against their pay-back periods. The purpose of this research was to integrate the iterative process into a high level of integrative process approach at an early stage of the project (Feasibility and Programming stage). The aim was to concentrate on the LEED categories that contribute more to the project in terms of point allocations without draining the project's budget at a very early stage of the project.

El Rio: A Student Research Journal. Vol. 3, No. 1 (2020), pp. 3–13.

Introduction

Since LEED (BD+C) rating system consists of seven categories according to the score-card of New Construction and Major Renovation, this paper proposes the utilization of new ideas and materials that could be used for commercial buildings (e.g. educational building). These new ideas and materials help to accomplish synergy between credit category, system, and components that can be realized through the integrative process to achieve high levels of building performance, human well-being, and environmental benefits. Each LEED credit is assigned points based on its contribution toward address-ing one or more of the LEED impact categories. The credit categories are composed of required and optional green building strategies. Required strategies are referred to as "pre-requisites." Optional strategies are referred to as "credits." To receive certification, projects must achieve all prerequisites and a minimum of 40 out of the available 110 points. Higher levels of certification are achieved by earning more points: LEED Certified is 40–49 points, LEED Silver is 50–59 points, LEED Gold 60–79 points, and LEED Platinum is 80 or more points. Proactive project teams typically target 3–4 more points than the minimum number of points necessary to achieve the targeted certifica-tion level. This is an effective risk management strategy since it is possible that a few credits may no longer be feasibly pursued during the design and construction phases of a project or that credits will be denied during the Green Business Certification Inc. (GBCI) review.

Typically, educational buildings are built or renovated using conventional materials (e.g. concrete and steel), which are not useful in terms of minimizing the embodied energy and other impacts associated with the extraction, processing, transport, maintenance, and disposal of building materials (Alshareef 2018). Therefore, this paper shifts the focus toward more eco-friendly materials that are not typically utilized in commercial buildings. Not only are these materials are feasible, efficient, and have a less negative impact on the environment, but also they achieve all prerequisites and credit points. As a result, stakeholders and decision-makers' participation at a very early stage of the project are highly recommended.

Methodology

The purpose of this case study is to explore different approaches to improve the efficiency of potentially newly renovated buildings (i.e. Technology Building at CSU-P). The paper collaborators began with a general observation of the Technology Building's needs with the LEED point system in mind. This approach is used because it produces generalized concepts and decisions based on a small number of observations. This is happened by iteratively meeting daily, weekly, and monthly to refine the objectives and goals and to find sustainable solutions. Also, several personal interviews were conducted with stakeholders and owner representatives to integrate their ideas and visions into this research and refine the overall processes. Furthermore, a high level of an integrative process was conducted through three phases of evaluations, such as Discovery, Implementation, and Occupancy phases. For this research, data was collected mostly by interviewing construction faculty members, owner representatives, designers, general contractors, and the paper contributor's personal observations. Afterward, all the col-laborative approaches are examined against the LEED category sections concerning the prerequisites and credit conditions.

Case Study-Technology Building at CSU-Pueblo

Materials and Resources

Trash and Recycling Bins

This project will need an adequate number of dumpsters and recycling units to hold the amount of trash and any other type of material that is being removed while this building is under construction. For this project, the dumpster that should be used for the job is a 40-yard construction dumpster. A 40-yard dumpster typically includes a 5–6 tons or 10,000–12,000 pound weight limit, though weight limits vary by location and type of disposed of material. For the size of the building, there should be 4 dumpsters that are required to hold all the material that is to be removed. The size and number of dumpsters are prerequisites under the Materials and Resources category and no points are collected. A suitable location to store and sort all the materials that are being installed or being removed to reuse later on is a significant step in the construction process, which contributes to the prerequisites of the Construction and Demolition Waste Management Planning category. The best location to stage everything is going to be in the parking areas behind the Technology Building, parking lot S-1 (shown in Fig. 1). This is the ideal location because it is large enough to store all of the building materials that are needed and it is convenient for trucks or equipment to enter and exit the job site. The authors recommend that the best location have everything staged for this project to be in the S-1 parking lot behind the Technology building. This gives the best location to have materials delivered and to have the dumpster placed in a good location that will make it easy to dispose of the waste that is being taken out of the building. And by making this location the designated area for staging, the Storage and Collection of Recyclables prerequisites can be achieved.

Exteriors

Exterior Insulation Finishing System (EIFS)   EIFS as an exterior provides insulation of an R-value of up to R20. It improves the energy efficiency of the building envelope and is environmentally friendly. EIFS helps to achieve partial Optimize Energy Performance credit (16 points). Unlike wood, siding, stucco, and other siding materials, EIFS rarely need painting and are highly durable year-round, and they are even capable of withstanding powerful hurricanes. To move 25,000 square feet of material, EIFS require

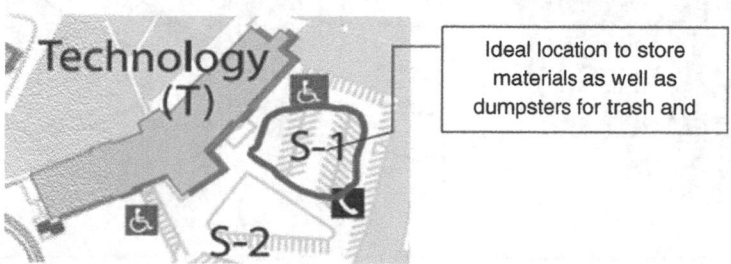

FIGURE 1   South S-1 parking lot, CSU-Pueblo map.

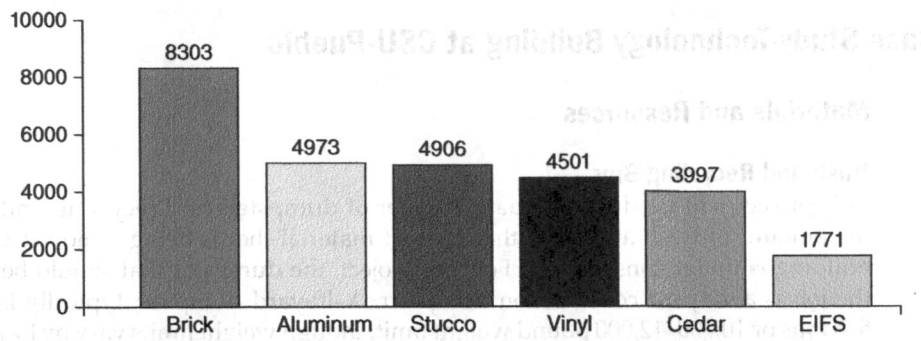

FIGURE 2 EIFS has a significantly smaller carbon footprint when compared to other materials.

16 times fewer tractor trucks than bricks and 6 times less than stucco (Wasmi 2016). EIFS also support sustainable design practices such as achieving LEED Building Certifications. This material does great with managing air and moisture infiltration as well as condensation. EIFS has been proven to produce the smallest carbon footprint of all other claddings according to the National Institute of Standards and Technology (shown in Fig. 2).

All EIFS panels include a fluid-applied water-resistive barrier coating that is applied to the exterior face of the structure, and exterior insulation is adhesively attached using a notched trowel to provide vertical paths for water drainage. Next, a base coat, either acrylic or polymer-based cement material, is applied to the top of the insulation and then reinforced with glass fiber reinforcement mesh. The reinforcement mesh is embedded in the base-coat material. The finish is a textured coat that's decorative and protective, as shown in Fig. 3.

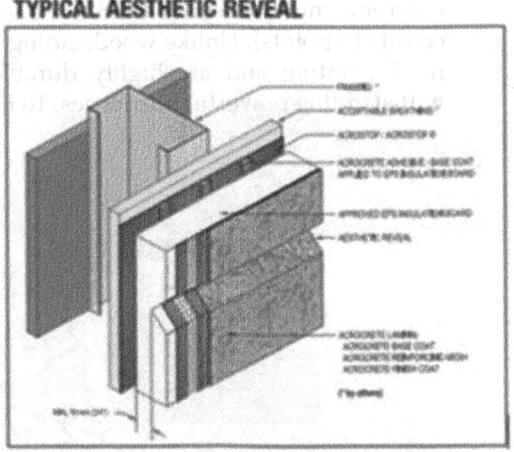

FIGURE 3   Two most common EIFS applications on exteriors.

Exterior Options	Total Building Square Footage (ft²)	Material Cost Per Footage ($)	Total Estimated Cost of Material ($)
EIFS	24,900	$16	$398,400
Stucco	24,900	$20	$498,000
Brick	24,900	$22	$547,800

TABLE 1 Total Estimated Cost of Material Using Different Exterior Applications

The main drawbacks with EIFS (Lstiburek 2007) are that they need to be recoated every 10 years to maintain the system, and if not properly installed, it can create problems that result in leakage. However, these installation problems can easily be avoided. A proper EIFS installation will shed water and be sealed at the windows and other wall penetrations so that leakage doesn't occur. Given the facts and information presented on EIFS, the authors recommend that the Technology Building chooses to use this highly efficient exterior system to achieve LEED certification for the Technology Building. With this suggestion, there is a rough estimate of the EIFS material cost (Canova 2013) compared to other commonly used materials (shown in Table 1).

By running takeoffs using Blue Beam (see Fig. 4) the authors were able to calculate the total area of the exterior, which was then multiplied by the cost of material per square foot, which then gave the proposed total of $398,000 to use EIFS on the entire Technology Building. This price includes material transportation and labor costs.

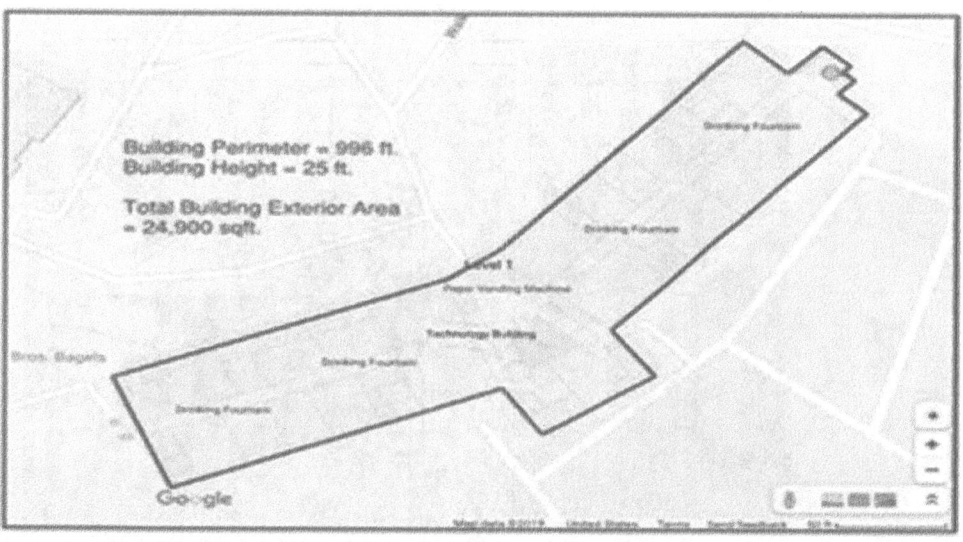

FIGURE 4 Takeoffs of the exterior on Technology Building using Blue Beam.

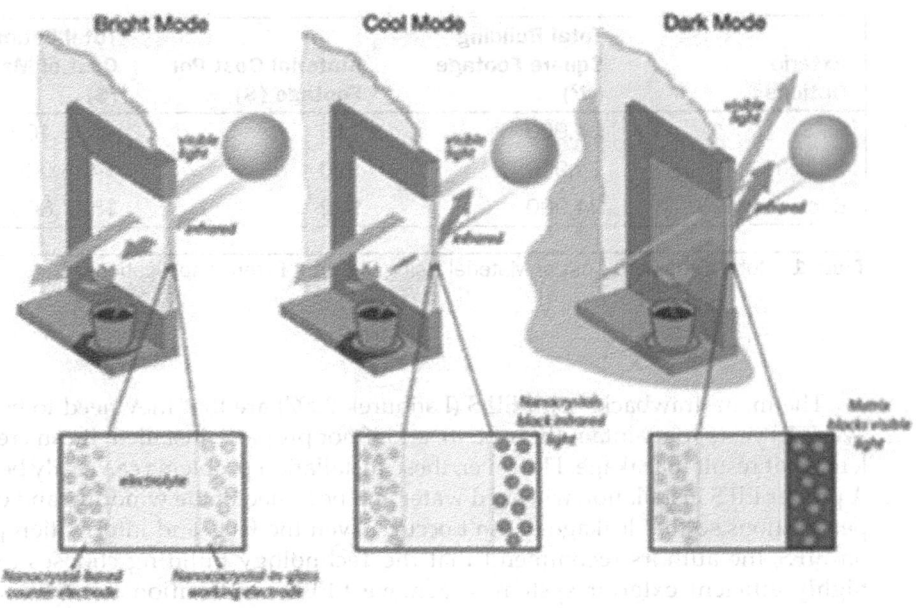

FIGURE 5   Smart Glass Technology effects on windows when the light transmission is altered.

Windows

Smart Glass Technology is a glass or glazed window whose light transmission properties are altered when voltage, light, or heat is applied. When a small electrical current is applied to the ceramic layers coated on the glass, it causes lithium ions to change layers, which causes the glass to tint. Reversing this polarity causes the glass to clear. Simply turn a dial to apply a small amount of electricity to the glass, and it will absorb infrared light. Turn another dial, and the glass will go dark (shown in Figs. 5 and 6).

FIGURE 6   Before and after examples of Electrochromic Smart Glass in use.

The effects are substantial when using electrochromic (Lee 2007) windows. Being that the Technology Building is sitting on top of a hill overlooking Pueblo, it is the authors' recommendation to use such technology because of the long hours of sunlight the college receives throughout the year, especially during the long hot summers. While significant energy savings is a big reason to consider smart glass for the Technology building, it isn't the only benefit. Smart glass gives building occupants a connection to the natural outdoor environment by providing quality views without having to sacrifice any scenery. The openness the smart glass creates in an office environment promotes happiness, creativity, and communication, which are important aspects of good design and sustainable building. Other advantages of electrochromic windows are that it brings the heat load of the building down (Somani 2002). According to scientists at the US Department of Energy's National Renewable Energy Laboratory (NREL), windows like this could save up to 8 percent of a building's total energy consumption. These windows use only tiny amounts of electricity to switch from dark to light which translates to huge net savings overall. With that said, HVAC systems can also be smaller, reducing overall capital expenses. Improved thermal comfort and a reduction of glare for the building's occupants can be achieved.

Smart glass was designed to maximize the use of natural daylight in buildings to improve the well-being of the people within. The use of smart glass on such a commercial project can help CSU-Pueblo achieve up to 37 LEED Certification Credits (chart is shown in Fig. 7). Part of those credits being one for innovative technology. By incorporating this design strategy that enhances daylight penetration with the use of smart windows, designers can additionally increase the number of occupants with exterior views. Adding smart windows and sidelights to openings built with this technology can help assist this project move forward in achieving this LEED Certification.

INTUS WINDOWS
LEED v4 SCORECHART

LEED Category	LEED Credit	Points
Energy & Atmosphere	Optimize Energy Performance	18
Materials & Resources	Building Product Disclosure & Optimization - Product Declaration	2
	Building Product Disclosure & Optimization - Material Ingredients	2
Indoor Environmental Quality	Enhanced Indoor Air Quality Strategies	3
	Indoor Air Quality Assessment	2
	Daylight	3
	Quality Views	1
	Acoustic Performance	1
Innovation	Innovation	5
Grand Total		Up to 37

* max LEED points available

Figure 7   LEED v4 Score Chart showing a total of 37 points applicable to the use of Smart Glass Technology.

Water Efficiency

Indoor Water Usage

The Technology building has four restrooms, two men and two women. In each woman's bathroom, there are four sinks and five toilets. In each men's bathroom, there are 4 sinks, 3 toilets, and 3 urinals. There is a total of 16 sinks, 16 toilets, and 6 urinals. All plumbing fixtures should be WaterSense labeled or similar because WaterSense label is an EPA standard for water efficiency. The U.S. Green Building Council (USGBC), states the most often used fixtures are high-efficiency toilets and non-water urinals. If there are lower reduction needs of 20% to 30%, dual flush and high-efficiency urinals were most often selected. The fixtures to evaluate were selected out of the 2019 Wholesalers List Price Book. The toilet, urinal, and faucet models selected are WETS-2002.1201 with Sloan ECOS 8111, WEUS-1000.1201 with SOLIS 8186, and EBF 615, respectively. The EBF 615 is a battery-operated model and is to be assessed with an electric hardwired model, ETF-80.

Outdoor Water Usage

Outdoor water reduction was considered with the methods of xeriscaping and little or no irrigation. Xeriscaping is an approach that includes efficient irrigation and native plant species to reduce outdoor water usage. Native plant species reduce outdoor water use because the species is already adapted to the climate, therefore, there is no need to water the plants. Irrigation was approached by comparing drip irrigation with the conventional sprinkling system. Drip irrigation was considered because the system targets the roots of the vegetation directly and reduces the amount of runoff because over-spraying is avoided and it uses far less water than conventional sprinkler systems. A diagram of a drip irrigation system is shown in Fig. 8 provided by WP Law (2016).

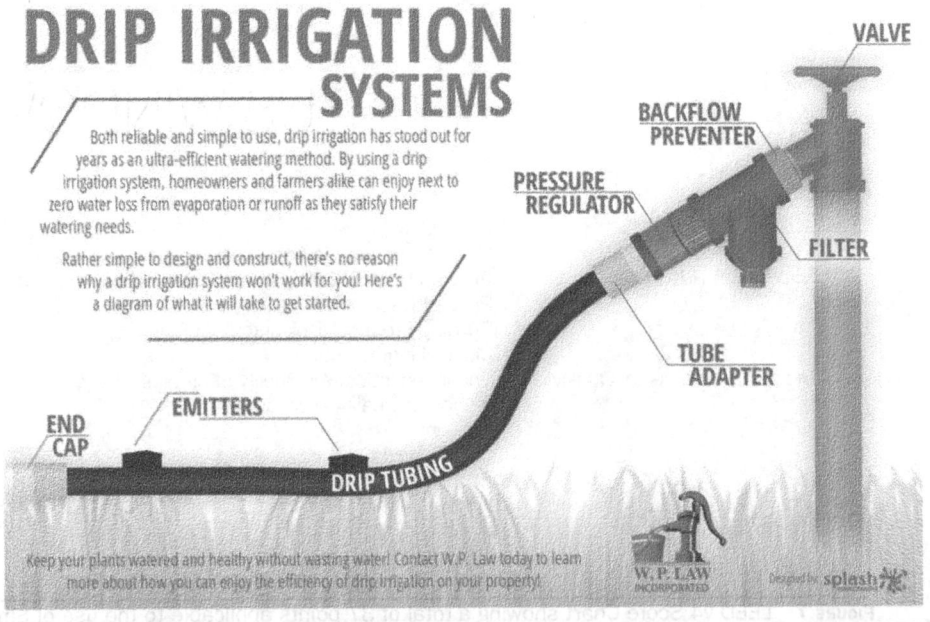

Figure 8 Drip irrigation systems, diagram by WP Law, 2016.

Water Usage

To reduce the amount of potable water, wastewater can be reused. The Technology Building does not have a system in place to reuse water. Treated wastewater can be used for toilet and urinal flushing. This will reduce the amount of potable water being used for purposes other than drinking. A type of wastewater treatment that was considered is the Living Machine. A living machine, according to Parsons Engineering Science (2001), will treat all wastewater so it can be used to flush toilets and urinals. If toilets and urinals are using reclaimed water to be flushed, that reduces the amount of water being used for purposes other than drinking. A diagram of Port of Portland's Living Machine is shown below in Fig. 9, showing the indoor and outdoor cells.

The U.S. EPA compared a living machine to a conventional wastewater system. The Present Worth Comparison of "Living Machines" and Conventional Systems table shows the cost of living machines with and without a greenhouse compared to a conventional wastewater system depending on the gallons used per day. Table 2 shows the Living

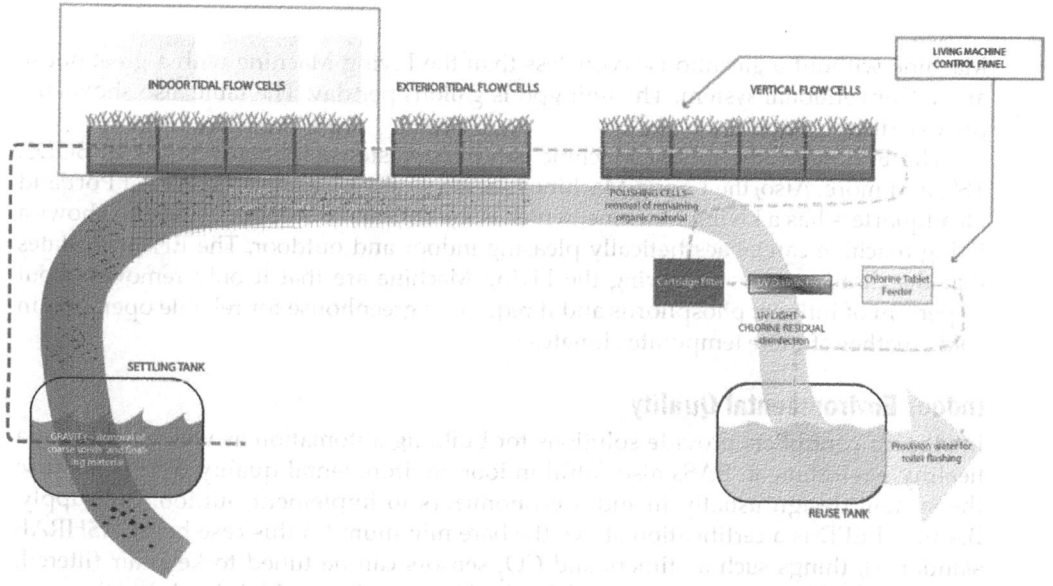

Figure 9 Port of Portland's living machine diagram by Yidan, Tamara, and Pure, 2016.

Process	40,000 gpd	80,000 gpd	1 million gpd
"Living Machine" with greenhouse	$1,077,777[1]	$1,710,280[1]	$10,457,542[2]
"Living Machine" without greenhouse	$985,391	$1,570,246	$9,232,257
Conventional System	$1,207,036[1]	$1,903,751[1]	$8,579,978[2]

(1) Cost difference is less than 20 percent.
(2) Cost difference is greater than 20 percent.
Source: U.S. EPA, 2001.

Table 2 Present Worth Comparison of "Living Machine" and Conventional Systems by U.S. EPA, 2001

FIGURE 10   Exterior (left) and interior (right) flow cells from Port of Portland case study, 2013.

Machine without a greenhouse costs less than the Living Machine with a greenhouse and a Conventional system. The unit gpd is gallons per day. The table also shows the price of different gpd.

The EPA states the Living Machine can treat wastewaters many needs to BOD5, TSS, and more. Also, the Living Machine is aesthetically pleasing. The Port of Portland Headquarters has a Living Machine, which is indoors and outdoors. Figure 10 shows a living machine can be aesthetically pleasing indoor and outdoor. The EPA also states that the disadvantages to having the Living Machine are that it only removes about 50 percent of influent phosphorus and it requires a greenhouse for reliable operation in cold weather of more temperate climates.

Indoor Environmental Quality

Intelligent controllers provide solutions for building automation as well as creating a healthy environment. BASs also fulfill indoor environmental quality credits because the system design usually includes economizers to implement outdoor air supply. Because LEED is a certification above the bare minimum (in this case being ASHRAE standards), things such as timers and CO_2 sensors can be tuned to keep air filtered, thereby eliminating the possibility of sick building syndrome. Lighting is another area of the building that contributes greatly to occupancy health, it also contributes greatly to power consumption. In fact, LEED cited lighting power as the largest electricity consumer in commercial buildings (LEED GA v4.1). Remedies to excessive power consumption in buildings result in addressable light controls, changes to color temperature, and regulated daylight.

Findings

Battery vs. Electric

Touchless faucets are useful for conserving water. Because the water shuts off automatically, this can dramatically cut down on water waste and reduce the risk of sink overflow caused by the faucet being left on. Some benefits to the electric hardwired faucet are that sensor taps would pour 10 L to 15 L per minute, while the other type would not use more than 6 L. This practice can benefit the environment more if the power comes from a renewable source and if further restraints are made in the system to control the out-flow of water. A disadvantage to the electric operation is that if the building loses power, it can cause the faucets to stop working. Battery-powered faucets are beneficial because both are the same as the electric ones in every way in saving water consumption and effi-ciency. Another disadvantage to the battery-powered is that batteries are replaceable often. The battery-powered is good because if the power happens to go out, the faucets will continue to work. The most common one used for buildings like this is the electric type. The electric faucet is the most commonly used because both are easy to install and there is hardly any maintenance required. Even though each option is useful, the elec-tric faucet is the best option for the job. Therefore, the EBF 615 battery-powered is not considered, instead, the ETF-80 hardwired model will be considered.

Water Efficiency

Indoor Water

The Water Resource category will affect the overall use of potable water in the Technology Building. The authors' final decision on water efficiency for indoor use is to use efficient fixtures for the toilets, urinals, and faucets. The total number of fixtures are shown in Table 3. The total number of fixtures were found by counting each fixture and recorded in a spreadsheet.

The cost of the fixtures based on the list price is shown in Table 4. The total cost for all fixtures is found by multiplying the total amount of each fixture by the List Price found in the Wholesalers List Price Book.

	Men	**Women**	**Total**
Number of Bathrooms:	2	2	4
# of Sinks:	4	4	16
# of Toilets:	3	5	16
# of Urinals:	3	0	6

TABLE 3   Number of Fixtures, Table Made in Excel

Efficiency	**Model**	**List Price:**	**Cost for All Fixtures**
1.6/1.1 gpf	WETS-2002.1201 w/Sloan ECOS 8111	$817.15	$13,074.40
0.125 gpf	WEUS-1000.1201 w/SOLIS® 8186	$958.95	$5,753.70
0.35 GPM	ETF-80	$752.65	$12,042.40
		Total	$30,870.50

TABLE 4   List Cost of Fixtures Made in Excel Spreadsheet

	Conventional Sprinkler System		Drip Irrigation System	
	Low	**High**	**Low**	**High**
Initial Cost	$1,540.00	$2,240.00	$1,707.00	$2,029.89
Annual Cost over Lifetime	$1,371.73	$1,408.93	$445.95	$504.77
Life Time (year)	20	12	25	20
Total Cost	$28,974.60	$19,147.16	$12,855.75	$12,125.29

TABLE 5 Irrigation Comparison Made in Excel Spreadsheet

Outdoor Water

The decision on outdoor water use is to focus on more efficient irrigation and landscaping by using native plants, and drip irrigation to avoid excessive use from runoff. The EPA has a study on the types of irrigation, which states that a conventional sprinkler system has an initial cost, low and high, of $1,540.00 and $2,240.00, respectively. The total annual costs over its low lifespan of 20 years and high lifespan of 12 years are $1,371.73 and $1,408.93, respectively. A sub-surface drip irrigation system has an initial cost, low and high, of $1,707.00 and $2,029.89, respectively. The total annual costs over its low lifespan of 25 years and high lifespan of 20 years are $445.95 and $504.77, respectively.

The drip irrigation system may have a greater initial cost; however, over its lifetime, there is a lowercost annual cost which makes the drip irrigation system's total cost cheaper than the conventional sprinkler system. The data and results are shown in Table 5. The Initial Cost, Annual Cost over Lifetime, and Life Time are from the EPA case study.

Water Usage

The final decision for the Water Resource category is to use The Living Machine to reduce the amount of potable water consumption by using reclaimed wastewater for flushing toilets and urinals. Based on the Port of Portland Headquarters' case study on the Living Machine, the system capacity is 5,000 gpd of a 200,000 square

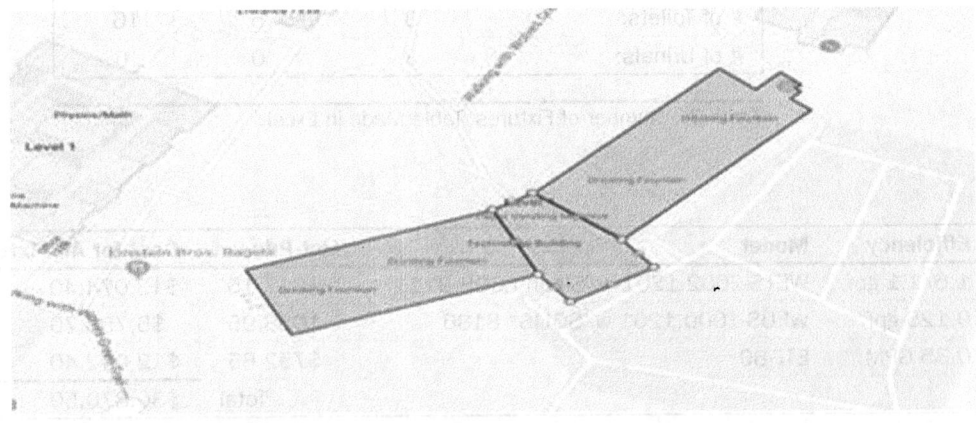

FIGURE 11 Technology outline from mapdevelopers.com.

foot headquarters building. The area of the Technology Building, which is about 29,600 square foot was found by using the website Mapdevelopers' area finder (see Figure 11). By comparing the Port of Portland's system capacity and size of the building to the Technology Building, a living machine of similar criteria will support the size of the Technology building with ease. If the price of a 40,000 gpd system is $1,077,777, the worth of the Living Machine for the technology building could be less than the worth of a Living Machine with a greenhouse.

A Living Machine needs a greenhouse when the climate is too cold, which is why the Technology Building should install a Living Machine similar to the Port of Portland Headquarters interior and exterior Living Machine. Having the fragile part of the Living Machine inside of the building removes the need for a greenhouse.

Conclusion

This research considered synergy at a very early stage of projects (i.e. Feasibility and Programming stage) to encourage stakeholders and decision-makers' involvements to have a more collaborative environment with fewer budget constraints. Typically, changes at this stage are expected and have fewer impacts on the budget; more so, participants are more resilient to new ideas. Therefore, this research studied new ideas and materials that are rarely utilized in commercial buildings (i.e. educational buildings) to develop more eco-friendly and sustainable buildings. The effort will lead to transforming the way buildings and communities are designed, built, and operated, enabling an environmentally and socially responsible, healthy, and prosperous environment that improves the quality of life.

The research is used to gain a better insight into the possibilities for improvement of the Technology Building at CSU-Pueblo as an educational case study building. It was approached to produce generalized concepts and conclusions by gathering documents and reports for the most useful materials, items, and ideas to compare them to their pay-back periods. The LEED categories were the vehicle to drive the research study to evaluate the tradeoff of materials against credit points while considering the project's budget at an early stage. Furthermore, multiple iterative processes were incorporated into a high level of integrative process to refine the outcome and collaborate with subject matter experts. This approach helped to provide a more efficient methodology and eliminated unnecessary overlaps and inefficiencies.

References

Canova, D. (2013, April 8). Costs of EIFS systems. (J. Dentz, Interviewer)

Lstiburek, J. (2007). Building Science Digest 146: EIFS—Problems and Solutions. Sommerville, MA: Building Science Press.

Lee, E.S., Yazdanian, M. & Selkowitz, S.E. (2004). The Energy-Savings Potential of Electrochromic Windows in the U.S. Commercial Buildings Sector. LBNL-54966 Lawrence Berkeley National Laboratory, Berkeley.

National Renewable Energy Laboratory. (n.d.). National Residential Energy Efficiency Measures Database. Retrieved May 13, 2013, from National Renewable Energy Laboratory: http://www.nrel.gov/ap/retrofits/measures. cfm?gId=12&ctId=410&scId=6547.

Parsons Engineering Science, Inc. and Environmental Engineering Consultants. (2001). Performance Comparison: The Burlington and Frederick Living Machines. In EPA, The Living Machine (R) Wastewater Treatment Technology : An Evaluation of Performance and System Cost (pp. 61–80).

Somani, P.R. & Radhakrishnan, S. (2002). Electrochromic materials and devices: present and future. Materials Chemistry and Physics 77, 117–133.

Sloan Valve Company. (2019, September 29). 2019 Wholesalers List Price Book. Retrieved from: https://www.sloan.com/sites/default/files/2019-09/Sloan_Master_Price_Book_0.pdf.

WP Law, (2016, March 31). Drip Irrigation System Breakdown. Retrieved from: http://wplawinc.com/landscape-irrigationblog?Tag=irrigation_design&PageID=.

Alshareef, H. A. (2018). Initial analytical investigation of overhead sign trusses with respect to remaining fatigue life and predictive methods for inspection (Doctoral dissertation).

Wasmi, H. A., & Castro-Lacouture, D. (2016). Potential impacts of BIM-based cost estimating in conceptual building design: a university building renovation case study. In Construction Research Congress 2016 (pp. 408–417).

LEED Process Assessments and Efficiency Improvements for Renovated Buildings

Husam A. Alshareef, Anthony Clark, Alexander Milyard, Hamern Robert, Brian Hund

Abstract

The Library and Academic Resources Center (LARC) at Colorado State University—Pueblo (CSU-P) was renovated in 2011. During this time, the building was awarded Leadership in Energy and Environmental Design (LEED) Platinum. This is the highest award for a sustainable building granted by the United States Green Building Council (USGBC). This building was evaluated under the LEED version 2.2 Building and New Construction standard. The LARC building is studied and evaluated in this research as a case study. All three LEED phases were evaluated during this case study: discovery, implementation, and occupancy. The purpose of this case study is to assess the LEED process used during the first evaluation and propose any necessary improvements to increase the efficiency of the building. The secondary purpose was to determine if the building could achieve a lower LEED award without compromising efficiency. This study was conducted by interviewing campus LEED professionals, observing LEED literature in the LARC building, and utilizing publicly available information. Our analysis results in a proposal that increases the LEED score to 57 out of 69 points for an award of LEED Platinum. The infrastructure proposed in this paper could lead to an increased LEED score for all buildings on campus.

Introduction

This report provides a comprehensive study and analysis of the Colorado State University—Pueblo LARC building. The study focuses on the way in which the building was certified as a LEED Platinum project. Aspects in the LEED process range from the physical materials used during the project, to environmental quality during construction and during the occupancy of the building after completion (Wasmi 2016). LEED certification also looks at the interior and exterior of the structure, such as indoor environmental quality, energy and atmosphere, and water efficiency; all of these topics were investigated throughout the duration of the case study. After the initial findings, it was found that the LARC building—even with a Platinum rating—could attain a handful of other LEED credits by using different methods and/or materials to increase the sustainability and efficiency of the building.

El Rio: A Student Research Journal. Vol. 3, No. 1 (2018), pp. 14–21.

Methodology

This case study followed a three-stage or phase methodology that tried to emulate the actual process in which a LEED project is developed and implemented. The first phase of our investigative methodology is the discovery phase. At this time in the study, the investigative team does not yet know what different means the builders used to achieve certification in LEED. In the discovery phase, the investigative team's goal was to begin to understand the project as a whole and then slowly break the scope of the project up into smaller components. These smaller components then became focal points for the rest of the case study. After understanding the focal points of the project, the discovery began to analyze the way in which the project achieved their LEED Platinum certification. Due to the fact that this case study is based around an existing building, the different systems, means, and methods used to reach said certification were relatively easy to find and understand. The last part of the discovery phase was to locate certain LEED credits that were not awarded to the LARC building and possible reasons why they were not achieved. It was also in this part of the discovery phase in which the investigators could begin to plan for the next part of the phasing, the implementation phase. During the implementation phase of the study, the investigators were to use the information found in the discovery phase and find areas that the LARC building could improve on and possibly earn more credits toward their certification. Such areas include sustainable sites, water efficiency, energy and atmosphere, materials and resources, and indoor environmental quality. Each LEED-accredited area was investigated in depth during the implementation phase in order to produce a better solution to the problems that the LARC could face based on their existing certification. This phase of the case study helped immensely due to the fact that it narrowed the focus of the investigation down to a few key concepts that could be improved on. It also allowed the investigators the chance to further research the different methods used to mitigate certain problems that other LEED projects had faced. The last phase of the methodology was the occupancy phase. This phase took the different methods recommended in the implementation phase and discussed whether or not these methods would be beneficial to the LARC building. This phase looked at the cost of each introduced method, what each method requires, both positive and negative aspects of each, and the rate of return on investment if there is any at all. This final phase provided the best results for recommendations to be made at the end of the case study.

Case Study—LARC Building at CSU-Pueblo

The LARC project achieved LEED Platinum certification with version 2.2 The renovation was considered a major renovation because the building's acoustics, exterior aesthetics, day lighting, and overall Mechanical, Electrical, and Plumbing systems (MEP) were improved. The H.W. Houston was the general contractor for the renovation of the Library and Academic Resource Center (LARC) project. The architect firm was Bennett Wagner and Grody. The total square footage of the building was 125,800.

There are many alternatives to the LEED building design that do exist according to the Harvard Energy and Facilities committee. Throughout this research, it was discovered that it is no secret that creating an energy-efficient building is quite a desirable goal. Building green saves on energy and waste costs and limits the negative impact on the environment (Alshareef 2018). There are a lot of alternatives that are available and would need to be considered before committing to an endeavor such as this.

The initial investment in becoming LEED certified can be quite significant. It was discovered that this needs to be kept in consideration. Becoming LEED certified is not only a complicated process, it's also expensive. There is a flat registration fee ranging from $1,200 for the basic certification to $3,250 for silver, gold, and platinum certification, which is just for the precertification review. There are also additional costs, depending on the size of the building, and these costs can reach up to $27,500 for buildings with more than 500,000 square feet. Moreover, the things that must be changed in an existing structure to achieve the certification can cost hundreds of thousands of dollars. It was discovered that there is not only a large monetary investment involved, but that there is also a significant time investment to achieve LEED certification. Further, and upon becoming LEED certified, there is an investment in maintenance as well.

Water Efficiency

There are two alternatives that CSU-P should consider in order to increase water efficiency on campus: non-potable water usage or xeriscaping. These two ideas are not mutually exclusive; however, the cost to integrate a non-potable water irrigation system would be very high. Therefore, xeriscaping should be reduced if a non-potable system is installed.

Federal Energy Management program defines Xeriscaping as a landscape design practice that reduces or eliminates the need for irrigation. This is done by drastically reducing the surface area of the vegetative landscape. Often rocks or mulch are used, along with drought-tolerant plants. The main advantages of xeriscaping are that it greatly reduces water consumption, and it reduces maintenance and usage costs. However, grass is iconic on a college campus because it provides an environment in which students can come together. Therefore, a campus with a rock and mulch landscape may not be appealing to students.

Non-potable water is water that is not safe for human consumption, but it can still be used for other purposes. Non-potable water is highly effective for irrigation because it is cheaper than potable water, and vegetation can survive on it. There are three main ways of receiving non-potable water on campus: reclaimed water from a waste treatment facility, pumping the water directly from a water source, or collecting the water from a runoff on campus.

The James R. DiIorio Water Reclamation Facility treats wastewater in the city of Pueblo. After the water is treated, it is pumped directly into the Arkansas River. This water could be used more efficiently if it was used for irrigation. Unfortunately, the waste treatment facility is nearly 3.5 miles from the CSU-P campus, as seen in Fig. 1, and it would be very expensive to construct the pipe necessary to transport the water. Additionally, many businesses and residents would be impacted from the construction process.

Fountain Creek is approximately one mile from campus, and 1.25 miles from the LARC building, as shown in Fig. 2. Non-potable water could be pumped directly from Fountain creek; however, water rights would need to be obtained in order to do this. Although, it would be cheaper to construct the infrastructure necessary to pump water from Fountain Creek than from the water reclamation facility. A third option would be to construct a basin on the CSU-P campus that collects runoff. However, since CSU-P is on a hill, the basin would only get runoff from the campus itself. It is unlikely that the runoff from campus would be able to supply the entire campus's irrigation needs.

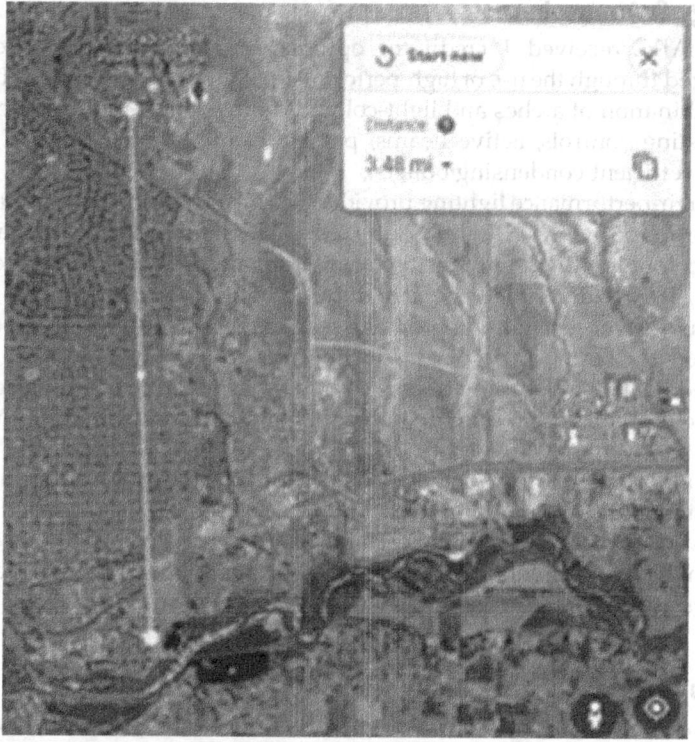

FIGURE 1   Distance from water reclamation facility to CSU-P.

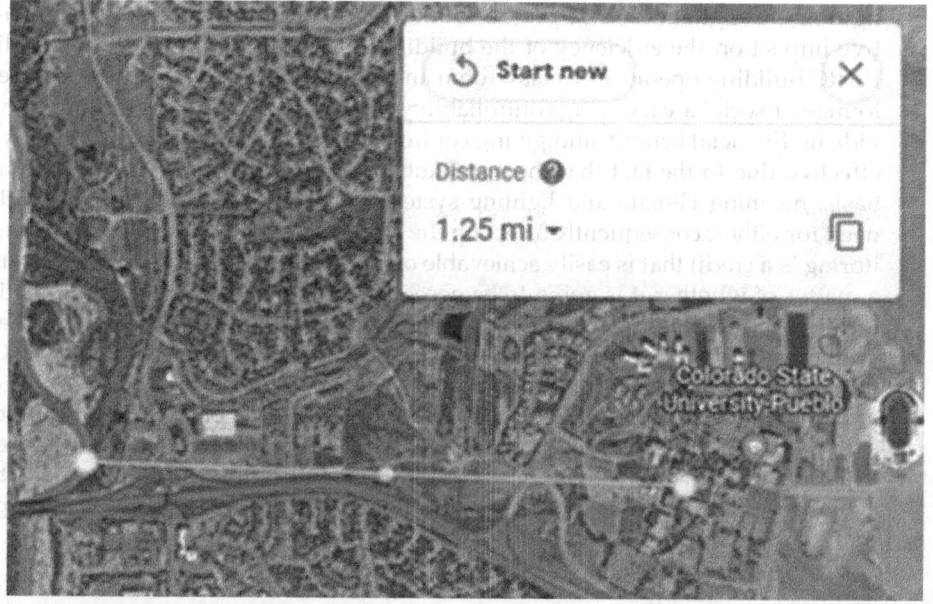

FIGURE 2   Distance from Fountain Creek to CSU-P.

Energy & Atmosphere

The LARC received 1 credit for optimizing energy performance. This credit was achieved through the use of high-performance lighting, T5HO lamps, LED accent lighting, a combination of arches and light-colored ceilings, high-performance glazing, daylight harvesting controls, active beams, passive beams, modern HVAC technologies, and highly efficient condensing boilers.

High-performance lighting provided warm and comfortable lighting very efficiently. The primary overhead lighting utilized T5HO lamps. The combination of arches and light-colored ceilings, with the type of lighting fixtures that shines the light up as well as down, provided evenly dispersed, diffused lighting that furnish less glare for computer screens. LED accent lighting provided new shapes of lighting. High-performance glazing allowed for larger areas of glass while preventing unwanted heat from entering or desired heat from escaping from the building. Daylight harvesting controls used photo-cells to monitor the amount of ambient light and dim the lighting to appropriate levels when sufficient daylight exists. The active beams used high-velocity air to induce warm room air to move through chilled coils. Passive beams rely on the natural flow of warm air rising and cool air falling to silently cool the space. The modern HVAC technologies achieved 43.5% energy savings. This was achieved through the use of chilled beams, radiant heated slabs, and the displacement of air systems. The highly efficient condensing boilers produced hot water more efficiently than traditional boilers.

Indoor Environmental Quality

The LARC building achieved a score of 10 out of 15 possible credits (reference Table 1) in Indoor Environmental Quality. The credits that the project did not receive were air delivery monitoring, controllability of systems for both heating and lighting, and daylight and views. An initial alternative was to allow the controllability of heating and lighting throughout the layout of the building. This was proven to have more of a negative impact on the efficiency of the building due to the function of the building. The LARC building operates as a classroom and learning environment, a study area, and a lounge or social area. Giving controllability of climate and lighting systems would provide no financial benefit, and giving control to the occupants in the building is not cost effective due to the fact that the occupants do not reside in the building on a regular basis, meaning climate and lighting systems would be left in operation without the need for either; consequently affecting the efficiency of the building. Air delivery monitoring is a credit that is easily achievable on almost any project in today's industry, it is a matter of whether it is going to be necessary or not depending on the functions and location of the building. Air delivery monitoring can provide feedback to the climate system inside of a building. This in turn can improve the efficiency of the HVAC system in the building so that the system is not in use when it does not need to be. With daylight and views, the LARC building is an area where this use of natural lighting could be extremely useful, both for economic reasons and social reasons. This is one aspect of the project where the investigation is deciding to make improvements and attempt to achieve the two credits associated with daylight and views. Some positive points to make about daylight and views are as follows:

- Lower energy costs (HVAC)
- When controlled, natural lighting generates hardly any heat at all

- Overall energy savings can range from 15 to 40 percent
- Can have a positive impact on productivity and satisfaction of occupants

Negative points of daylight and views:

- Significant initial investment
- If not planned properly, using natural lighting can result in undesirable heat gains in the building
- Direct sunlight penetration in classrooms and offices often produce unpleasant glares

If planned and designed properly, a new daylight and views system could be beneficial to the LARC building, as well as providing two more credits to the overall LEED score applied to the building.

	Indoor Environmental Quality	Max	Obtained	Proposed
Prereq 1	Minimum IAQ Performance	Required	Required	Required
Prereq 2	Environmental Tobacco Smoke (ETS) Control	Required	Required	Required
Credit 1	Outdoor Air Delivery Monitoring	1	0	1
Credit 2	Increased Ventilation	1	1	1
Credit 3.1	Construction IAQ Management Plan, During Construction	1	1	1
Credit 3.2	Construction IAQ Management Plan, Before Occupancy	1	1	1
Credit 4.1	Low-Emitting Materials, Adhesives & Sealants	1	1	1
Credit 4.2	Low-Emitting Materials, Paints & Coatings	1	1	1
Credit 4.3	Low-Emitting Materials, Carpet Systems	1	1	1
Credit 4.4	Low-Emitting Materials, Composite Wood & Agrifiber Products	1	1	1
Credit 5	Indoor Chemical & Pollutant Source Control	1	1	1
Credit 6.1	Controllability of Systems, Lighting	1	0	0
Credit 6.2	Controllability of Systems, Thermal Comfort	1	0	0
Credit 7.1	Thermal Comfort, Design	1	1	1
Credit 7.2	Thermal Comfort, Verification	1	1	1
Credit 8.1	Daylight & Views, Daylight 75% of Spaces	1	0	1
Credit 8.2	Daylight & Views, Views for 90% of Spaces	1	0	1
		15	10	13

TABLE 1 Indoor Environmental Quality LEED Credits (Scorecard of LARC Building)

Findings

It was further discovered by the investigators that in order for this building to strive closer to becoming a zero-point energy building, there would be many additional costs involved to achieve this. Specifically, the investigators had looked closely at incorporating a green, living roof to the building. It was therefore discovered that this could and would very likely cut down on the costs to supply food for this building's occupants. However, there is a significant financial investment the investigators found to implement this. The investigators did find that the overall level of self-sustainability could be improved with additional financial investment toward this goal.

Sustainable Sites

This first category of LEED Certification prerequisites has to do with the specific location and piece of land that the project is to be built on. It was discovered that these credits specifically deal with protecting the natural habitat in the area, keeping the open spaces open, dealing with the rainwater in the best way possible, and keeping the heat island effect and light pollution down to a minimum.

Site Assessment

This credit is worth 1 point. In order to earn this credit, project teams must perform and document a site assessment of the project location, including the following topics: topography, hydrology, climate, vegetation, soils, human use, and human health effects. The assessment should discuss how the topics above influence the design, as well as any of the topics that were not addressed in the design.

Protect or Restore Habitat

This credit is worth 1–2 points. The project must preserve and protect at least 40% of the greenfield (undeveloped) area on the project site, if such an area exists. In addition, the project must restore 30% of the site to natural habitat using native and adapted plant species (worth 2 credits) or provide financial support to an organization accredited by the Land Trust Alliance (worth 1 credit). The habitat restoration should include both soil and vegetation, and vegetated roofs can be counted in certain circumstances.

Open Space

This credit is worth 1 point. The project must provide open space equal to 30% of the total site area. At least 25% of that open space must be vegetated or have overhead vegetation. Turf grass areas do not count as vegetated areas. Open spaces must be designed for one or more of the following uses: social gathering, gardening, physical activity, or natural habitat that includes elements for human interaction. Vegetated roofs can be counted in certain circumstances.

Rainwater Management

This credit is worth 1–3 points. This credit asks the project team to design a rainwater management system that handles the water falling on the site in a way that is similar to the native state of the site. Depending on how much water the system is capable of handling, 1–3 points are possible. The capacity of the system is measured by what percentage of local or regional rain events could be handled by the system. If the system can handle 95% of the events, then it can earn 2 points, and 3 points for handling 98%. Or, as an alternate way of calculating the credit, if the system can handle 100% of the increase in runoff that occurs as the result of the development of the site from its natural state, then the project can earn 3 points.

Heat Island Reduction

This credit is worth 1–2 points. Heat islands occur in areas where hardscape surfaces (such as parking lots and sidewalks) hold heat and reflect it back, raising the temperature of the surrounding environment. This change in temperature can affect weather patterns in the local area. To avoid this, projects receive credit for using roofing materials with a high solar reflectance, reducing the number of hard surfaces, shading project areas with trees and other foliage, placing parking lots under cover, and using open paver systems.

Light Pollution Reduction

This credit is worth 1 point. Projects must reduce the amount of up-lighting used for exterior lighting, avoid pollution of light into adjoining sites, and control light levels outside the building to meet certain standards. This requires a photometric plan, which shows the level of light in all areas of the site. The design team must take measurements to confirm that the built condition meets the requirements for this credit.

Water Efficiency

Table 2, below, shows the points available in the water efficiency category. This table is a modified version of the 2013 LEED scorecard for the LARC. Five points can be obtained in this category, and CSU-P obtained three points. The research contributors believe that CSU-P should consider pursuing the "Water Efficient Landscaping" credit. In order to do this, a considerable investment would need to be made to bring non-potable water to the campus or remove irrigation altogether throughout the entire footprint of the LARC building.

Xeriscaping has a much cheaper total cost than installing non-potable water, but studies show that a grass environment is more appealing to humans. A xeriscaped environment may be detrimental to the recruiting efforts of the university; therefore, it is the recommendation of the research contributors that CSU-P investigates the feasibility of bringing non-potable water to campus. Bringing non-potable water to campus would

	Water Efficiency	Max	Obtained	Proposed
Credit 1.1	Water Efficient Landscaping, Reduce by 50%	1	1	1
Credit 1.2	Water Efficient Landscaping, No Potable Use or No Irrigation	1	0	1
Credit 2	Innovative Wastewater Technologies	1	0	0
Credit 3.1	Water Use Reduction, 20% Reduction	1	1	1
Credit 3.2	Water Use Reduction, 30% Reduction	1	1	1
		5	3	4

TABLE 2   Water Efficiency LEED Credits (Scorecard of LARC Building)

have a high upfront cost; however, the pipe network has a lifespan of 50–70 years. Once the infrastructure is in place it could be used for every facility on campus. This would give every structure on campus the water efficient landscaping credit. Additionally, it would considerably reduce irrigation costs and reduce the campuses potable water consumption.

Energy & Atmosphere

The alternate energy that could be used in the LARC building is a geothermal system. The benefits of a geothermal system can be configured to accommodate the amount of property used. A geothermal system can be configured to a horizontal or vertical, open or closed loop system. Horizontal loop systems have lower installation cost, but they require a plot of land sufficient for 3–5 trenches: 130 to 160 feet long and 12 to 20 feet apart. Water or antifreeze circulated through the pipes collects heat for heating in the winter and dumps heat for air conditioning in warm months. A vertical loop system has a higher installation cost, about $1500.00 per 12,000 BTUs (British Thermal Units). This system is ideal for smaller properties. Vertical loop systems are where several holes are drilled, each between 50–400 feet deep, and several pipes are installed. Water is then circulated through the pipes that collect heat for heating in the winter and dumps heat for air conditioning in warm months. The other benefits of this system are that they have a quiet operation, resulting in less noise pollution. Geothermal systems are more efficient than ordinary heating and air conditioning units because the systems deliver more energy than they use. A geothermal system will offer a more precise distribution of cooled or heated air, year-round, so there would be less hot and cold spots in the building. Geothermal technology is more reliable than most air conditioning units and heat pumps, and they typically require less maintenance than other heating and cooling units.

Heat pump pipes even have warranties of between 25 and 50 years, while the pump can usually last for at least 20 years. This also requires less space for hardware as opposed to conventional systems. This system is more environmentally friendly because geothermal systems don't emit carbon dioxide or other greenhouse gases that are considered contributors to environmental air pollution. This system is highly efficient because geothermal heat pump systems use 25% to 50% less electricity than conventional systems for heating or cooling, and with their flexible design, they can be adjusted to different situations, requiring less space for hardware as opposed to conventional systems.

Material & Resources

The LARC received 16.4 credits in the material and resources category. These credits varied from building reuse, recycled content, low-emitting material, and certified wood. The reason the LARC received 1.2 credits for building reuse was that it maintained 95% of the precast structure and cladding system. The exception was where panels were removed to allow for the expanded exterior glazing, which added more natural lighting. The reason the LARC received 4.2 credits for recycled content was because 20% of the building content was re-used. These materials varied from carpet, countertops, solar shades, and ceiling tiles. The reason the LARC received 4 credits for low-emitting material was because materials that have low-VOC (volatile organic compounds) content were used. These materials were adhesives, sealants, paints, coatings, and carpet systems. Composite woods and Agri fiber were selected to contain no urea-formaldehyde.

The reason the LARC received 7 credits for certified wood was because the wood that was selected were FSC (Forest Stewardship Council) certified.

Indoor Environmental Quality

This investigation's recommendation for improved Indoor Environmental Quality is to enhance the daylight and views in the LARC building. This comes as a recommendation due to its ability to save on energy and improve the social dynamic inside the building. Both of these focal points become more important based on the overall use of the building. The LARC is utilized as a classroom building, study area, and social/gathering area all at the same time. When natural daylight can improve productivity and satisfaction in these types of environments, the reason for this change is justified by the social improvement that it can have. The energy savings is more complicated. In order for the improved daylighting and views to be cost effective, it would be this investigation's recommendation to perform a design study before construction and/or improvements commenced. Such studies can be in the form of a Building Information Modeling (BIM) model, and when coupled with a specific location and time of year, the design team can resolve to the best solution possible.

Conclusion

Using the seven LEED (BD+C) rating system's categories, this report analyzed comprehensively the existing Platinum certification of an educational building (i.e. the LARC building at Colorado State University-Pueblo). This investigation was able to improve the overall score of the LEED accreditation by five points with respect to the budget. The intent of all explanations and recommendations is to ensure the betterment of the operation and sustainability of the LARC building, as well as to improve the building's LEED accreditation. The infrastructure proposed in this paper could lead to an increased LEED score for all buildings on campus (i.e. CSU-P), so this research serves as a vehicle for future investigation in this regard.

References

Alshareef, H. A. (2018). Initial analytical investigation of overhead sign trusses with respect to remaining fatigue life and predictive methods for inspection (Doctoral dissertation).

Federal Energy Management Program. (2009). Retrieved November 8, 2019, from https://www.energy.gov/sites/prod/files/2013/10/f3/water_fortcarson.pdf.

Harvard Energy & Facilities. (n.d.). Retrieved November 8, 2019, from http://www.energyandfacilities.harvard.edu/greenbuilding-resource/leed-case-studies.

Wasmi, H. A., & Castro-Lacouture, D. (2016). Potential impacts of BIM-based cost estimating in conceptual building design: a university building renovation case study. In Construction Research Congress 2016 (pp. 408–417).

APPENDIX B

LEED Scorecards

The LEED Scorecards help readers understand the credits pursued against prerequisites, and points earned through each adoption. Each adoption demonstrates emphasis over prerequisites and credits for each LEED category, so attention must be paid to the credit points allocation. There are eight adoptions listed in this appendix: Core and Shell, Data Centers, Healthcare, Hospitality, New Construction, Retail, Schools, Warehouse, and Distribution Centers. All the scorecards used in this appendix are referenced from the United States Green Building Council.

LEED v4.1 BD+C: Core and Shell
Project Checklist

Project Name:
Date:

Y	?	N			
			Credit	Integrative Process	1

			Location and Transportation		**20**
			Credit	LEED for Neighborhood Development Location	20
			Credit	Sensitive Land Protection	2
			Credit	High-Priority Site and Equitable Development	3
			Credit	Surrounding Density and Diverse Uses	6
			Credit	Access to Quality Transit	6
			Credit	Bicycle Facilities	1
			Credit	Reduced Parking Footprint	1
			Credit	Electric Vehicles	1

Y	?	N	**Sustainable Sites**		**11**
Y			Prereq	Construction Activity Pollution Prevention	Required
			Credit	Site Assessment	1
			Credit	Protect or Restore Habitat	2
			Credit	Open Space	1
			Credit	Rainwater Management	3
			Credit	Heat Island Reduction	2
			Credit	Light Pollution Reduction	1
			Credit	Tenant Design and Construction Guidelines	1

Y	?	N	**Water Efficiency**		**11**
Y			Prereq	Outdoor Water Use Reduction	Required
Y			Prereq	Indoor Water Use Reduction	Required
Y			Prereq	Building-Level Water Metering	Required
			Credit	Outdoor Water Use Reduction	3
			Credit	Indoor Water Use Reduction	4
			Credit	Optimize Process Water Use	3
			Credit	Water Metering	1

Y	?	N	**Energy and Atmosphere**		**33**
Y			Prereq	Fundamental Commissioning and Verification	Required
Y			Prereq	Minimum Energy Performance	Required
Y			Prereq	Building-Level Energy Metering	Required
Y			Prereq	Fundamental Refrigerant Management	Required
			Credit	Enhanced Commissioning	6
			Credit	Optimize Energy Performance	18
			Credit	Advanced Energy Metering	1
			Credit	Grid Harmonization	2
			Credit	Renewable Energy	5
			Credit	Enhanced Refrigerant Management	1

0	0	0	**Materials and Resources**		**14**
Y			Prereq	Storage and Collection of Recyclables	Required
			Credit	Building Life-Cycle Impact Reduction	6
			Credit	Environmental Product Declarations	2
			Credit	Sourcing of Raw Materials	2
			Credit	Material Ingredients	2
			Credit	Construction and Demolition Waste Management	2

0	0	0	**Indoor Environmental Quality**		**10**
Y			Prereq	Minimum Indoor Air Quality Performance	Required
Y			Prereq	Environmental Tobacco Smoke Control	Required
			Credit	Enhanced Indoor Air Quality Strategies	2
			Credit	Low-Emitting Materials	3
			Credit	Construction Indoor Air Quality Management Plan	1
			Credit	Daylight	3
			Credit	Quality Views	1

0	0	0	**Innovation**		**6**
			Credit	Innovation	5
			Credit	LEED-Accredited Professional	1

0	0	0	**Regional Priority**		**4**
			Credit	Regional Priority: Specific Credit	1
			Credit	Regional Priority: Specific Credit	1
			Credit	Regional Priority: Specific Credit	1
			Credit	Regional Priority: Specific Credit	1

0	0	0	**TOTALS**	**Possible Points:**	**110**

Certified: 40-49 points,   Silver: 50-59 points,   Gold: 60-79 points,
Platinum: 80-110 points

The scorecards were referenced from United States Green Building Council (USGBC)

LEED v4.1 BD+C: Data Centers
Project Checklist

Project Name:
Date:

Y	?	N			
0	0	0			1
			Credit	Integrative Process	1

Y	?	N		Location and Transportation	16
0	0	0		**Location and Transportation**	**16**
			Credit	LEED for Neighborhood Development Location	16
			Credit	Sensitive Land Protection	1
			Credit	High-Priority Site and Equitable Development	2
			Credit	Surrounding Density and Diverse Uses	5
			Credit	Access to Quality Transit	5
			Credit	Bicycle Facilities	1
			Credit	Reduced Parking Footprint	1
			Credit	Electric Vehicles	1

Y	?	N		Sustainable Sites	10
0	0	0		**Sustainable Sites**	**10**
Y			Prereq	Construction Activity Pollution Prevention	Required
			Credit	Site Assessment	1
			Credit	Protect or Restore Habitat	2
			Credit	Open Space	1
			Credit	Rainwater Management	3
			Credit	Heat Island Reduction	2
			Credit	Light Pollution Reduction	1

Y	?	N		Water Efficiency	11
0	0	0		**Water Efficiency**	**11**
Y			Prereq	Outdoor Water Use Reduction	Required
Y			Prereq	Indoor Water Use Reduction	Required
Y			Prereq	Building-Level Water Metering	Required
			Credit	Outdoor Water Use Reduction	2
			Credit	Indoor Water Use Reduction	6
			Credit	Optimize Process Water Use	2
			Credit	Water Metering	1

Y	?	N		Energy and Atmosphere	33
0	0	0		**Energy and Atmosphere**	**33**
Y			Prereq	Fundamental Commissioning and Verification	Required
Y			Prereq	Minimum Energy Performance	Required
Y			Prereq	Building-Level Energy Metering	Required
Y			Prereq	Fundamental Refrigerant Management	Required
			Credit	Enhanced Commissioning	6
			Credit	Optimize Energy Performance	18
			Credit	Advanced Energy Metering	1
			Credit	Grid Harmonization	2
			Credit	Renewable Energy	5
			Credit	Enhanced Refrigerant Management	1

Y	?	N		Materials and Resources	13
0	0	0		**Materials and Resources**	**13**
Y			Prereq	Storage and Collection of Recyclables	Required
			Credit	Building Life-Cycle Impact Reduction	5
			Credit	Environmental Product Declarations	2
			Credit	Sourcing of Raw Materials	2
			Credit	Material Ingredients	2
			Credit	Construction and Demolition Waste Management	2

Y	?	N		Indoor Environmental Quality	16
0	0	0		**Indoor Environmental Quality**	**16**
Y			Prereq	Minimum Indoor Air Quality Performance	Required
Y			Prereq	Environmental Tobacco Smoke Control	Required
			Credit	Enhanced Indoor Air Quality Strategies	2
			Credit	Low-Emitting Materials	3
			Credit	Construction Indoor Air Quality Management Plan	1
			Credit	Indoor Air Quality Assessment	2
			Credit	Thermal Comfort	1
			Credit	Interior Lighting	2
			Credit	Daylight	3
			Credit	Quality Views	1
			Credit	Acoustic Performance	1

Y	?	N		Innovation	6
0	0	0		**Innovation**	**6**
			Credit	Innovation	5
			Credit	LEED-Accredited Professional	1

Y	?	N		Regional Priority	4
0	0	0		**Regional Priority**	**4**
			Credit	Regional Priority: Specific Credit	1
			Credit	Regional Priority: Specific Credit	1
			Credit	Regional Priority: Specific Credit	1
			Credit	Regional Priority: Specific Credit	1

Y	?	N		TOTALS	Possible Points: 110
0	0	0		**TOTALS**	**Possible Points: 110**

Certified: 40-49 points,   **Silver:** 50-59 points,   **Gold:** 60-79 points,
Platinum: 80-110 points

The scorecards were referenced from United States Green Building Council (USGBC)

223

LEED v4.1 BD+C: Healthcare
Project Checklist

Project Name:
Date:

Y ? N

Y				
Y			Prereq	Integrative Project Planning and Design	Required
			Credit	Integrative Process	1

0	0	0	**Location and Transportation**		**9**
			Credit	LEED for Neighborhood Development Location	9
			Credit	Sensitive Land Protection	1
			Credit	High-Priority Site and Equitable Development	2
			Credit	Surrounding Density and Diverse Uses	1
			Credit	Access to Quality Transit	2
			Credit	Bicycle Facilities	1
			Credit	Reduced Parking Footprint	1
			Credit	Electric Vehicles	1

0	0	0	**Sustainable Sites**		**9**
Y			Prereq	Construction Activity Pollution Prevention	Required
Y			Prereq	Environmental Site Assessment	Required
			Credit	Site Assessment	1
			Credit	Protect or Restore Habitat	1
			Credit	Open Space	1
			Credit	Rainwater Management	2
			Credit	Heat Island Reduction	1
			Credit	Light Pollution Reduction	1
			Credit	Places of Respite	1
			Credit	Direct Exterior Access	1

0	0	0	**Water Efficiency**		**11**
Y			Prereq	Outdoor Water Use Reduction	Required
Y			Prereq	Indoor Water Use Reduction	Required
Y			Prereq	Building-Level Water Metering	Required
			Credit	Outdoor Water Use Reduction	1
			Credit	Indoor Water Use Reduction	7
			Credit	Optimize Process Water Use	2
			Credit	Water Metering	1

0	0	0	**Energy and Atmosphere**		**35**
Y			Prereq	Fundamental Commissioning and Verification	Required
Y			Prereq	Minimum Energy Performance	Required
Y			Prereq	Building-Level Energy Metering	Required
Y			Prereq	Fundamental Refrigerant Management	Required
			Credit	Enhanced Commissioning	6
			Credit	Optimize Energy Performance	20
			Credit	Advanced Energy Metering	1
			Credit	Grid Harmonization	2
			Credit	Renewable Energy	5
			Credit	Enhanced Refrigerant Management	1

0	0	0	**Materials and Resources**		**19**
Y			Prereq	Storage and Collection of Recyclables	Required
Y			Prereq	PBT Source Reduction- Mercury	Required
			Credit	Building Life-Cycle Impact Reduction	5
			Credit	Environmental Product Declarations	2
			Credit	Sourcing of Raw Materials	2
			Credit	Material Ingredients	2
			Credit	PBT Source Reduction- Mercury	1
			Credit	PBT Source Reduction- Lead, Cadmium, and Copper	2
			Credit	Furniture and Medical Furnishings	2
			Credit	Design for Flexibility	1
			Credit	Construction and Demolition Waste Management	2

0	0	0	**Indoor Environmental Quality**		**16**
Y			Prereq	Minimum Indoor Air Quality Performance	Required
Y			Prereq	Environmental Tobacco Smoke Control	Required
			Credit	Enhanced Indoor Air Quality Strategies	3
			Credit	Low-Emitting Materials	3
			Credit	Construction Indoor Air Quality Management Plan	1
			Credit	Indoor Air Quality Assessment	2
			Credit	Thermal Comfort	1
			Credit	Interior Lighting	1
			Credit	Daylight	2
			Credit	Quality Views	2
			Credit	Acoustic Performance	2

0	0	0	**Innovation**		**6**
			Credit	Innovation	5
			Credit	LEED-Accredited Professional	1

0	0	0	**Regional Priority**		**4**
			Credit	Regional Priority: Specific Credit	1
			Credit	Regional Priority: Specific Credit	1
			Credit	Regional Priority: Specific Credit	1
			Credit	Regional Priority: Specific Credit	1

0	0	0	**TOTALS**	Possible Points:	**110**

Certified: 40-49 points, **Silver:** 50-59 points, **Gold:** 60-79 points, **Platinum:** 80-110 points

The scorecards were referenced from United States Green Building Council (USGBC)

LEED v4.1 BD+C: Hospitality
Project Checklist

Project Name:
Date:

Y	?	N			
			Credit	Integrative Process	1

0	0	0	**Location and Transportation**		**16**
			Credit	LEED for Neighborhood Development Location	16
			Credit	Sensitive Land Protection	1
			Credit	High-Priority Site and Equitable Development	2
			Credit	Surrounding Density and Diverse Uses	5
			Credit	Access to Quality Transit	5
			Credit	Bicycle Facilities	1
			Credit	Reduced Parking Footprint	1
			Credit	Electric Vehicles	1

0	0	0	**Sustainable Sites**		**10**
Y			Prereq	Construction Activity Pollution Prevention	Required
			Credit	Site Assessment	1
			Credit	Protect or Restore Habitat	2
			Credit	Open Space	1
			Credit	Rainwater Management	3
			Credit	Heat Island Reduction	2
			Credit	Light Pollution Reduction	1

0	0	0	**Water Efficiency**		**11**
Y			Prereq	Outdoor Water Use Reduction	Required
Y			Prereq	Indoor Water Use Reduction	Required
Y			Prereq	Building-Level Water Metering	Required
			Credit	Outdoor Water Use Reduction	2
			Credit	Indoor Water Use Reduction	6
			Credit	Optimize Process Water Use	2
			Credit	Water Metering	1

0	0	0	**Energy and Atmosphere**		**33**
Y			Prereq	Fundamental Commissioning and Verification	Required
Y			Prereq	Minimum Energy Performance	Required
Y			Prereq	Building-Level Energy Metering	Required
Y			Prereq	Fundamental Refrigerant Management	Required
			Credit	Enhanced Commissioning	6
			Credit	Optimize Energy Performance	18
			Credit	Advanced Energy Metering	1
			Credit	Grid Harmonization	2
			Credit	Renewable Energy	5
			Credit	Enhanced Refrigerant Management	1

0	0	0	**Materials and Resources**		**13**
Y			Prereq	Storage and Collection of Recyclables	Required
			Credit	Building Life-Cycle Impact Reduction	5
			Credit	Environmental Product Declarations	2
			Credit	Sourcing of Raw Materials	2
			Credit	Material Ingredients	2
			Credit	Construction and Demolition Waste Management	2

0	0	0	**Indoor Environmental Quality**		**16**
Y			Prereq	Minimum Indoor Air Quality Performance	Required
Y			Prereq	Environmental Tobacco Smoke Control	Required
			Credit	Enhanced Indoor Air Quality Strategies	2
			Credit	Low-Emitting Materials	3
			Credit	Construction Indoor Air Quality Management Plan	1
			Credit	Indoor Air Quality Assessment	2
			Credit	Thermal Comfort	1
			Credit	Interior Lighting	2
			Credit	Daylight	3
			Credit	Quality Views	1
			Credit	Acoustic Performance	1

0	0	0	**Innovation**		**6**
			Credit	Innovation	5
			Credit	LEED-Accredited Professional	1

0	0	0	**Regional Priority**		**4**
			Credit	Regional Priority: Specific Credit	1
			Credit	Regional Priority: Specific Credit	1
			Credit	Regional Priority: Specific Credit	1
			Credit	Regional Priority: Specific Credit	1

0	0	0	**TOTALS**	Possible Points:	**110**

Certified: 40-49 points, **Silver:** 50-59 points, **Gold:** 60-79 points, **Platinum:** 80-110 points

The scorecards were referenced from United States Green Building Council (USGBC)

Project Name:
Date:

LEED v4.1 BD+C NEW CONSTRUCTION
Project Checklist

Y	?	N		
			Credit	Integrative Process	1

0	0	0	**Location and Transportation**	**16**
			Credit	LEED for Neighborhood Development Location	16
			Credit	Sensitive Land Protection	1
			Credit	High-Priority Site and Equitable Development	2
			Credit	Surrounding Density and Diverse Uses	5
			Credit	Access to Quality Transit	5
			Credit	Bicycle Facilities	1
			Credit	Reduced Parking Footprint	1
			Credit	Electric Vehicles	1

0	0	0	**Sustainable Sites**	**10**
Y			Prereq	Construction Activity Pollution Prevention	Required
			Credit	Site Assessment	1
			Credit	Protect or Restore Habitat	2
			Credit	Open Space	1
			Credit	Rainwater Management	3
			Credit	Heat Island Reduction	2
			Credit	Light Pollution Reduction	1

0	0	0	**Water Efficiency**	**11**
Y			Prereq	Outdoor Water Use Reduction	Required
Y			Prereq	Indoor Water Use Reduction	Required
Y			Prereq	Building-Level Water Metering	Required
			Credit	Outdoor Water Use Reduction	2
			Credit	Indoor Water Use Reduction	6
			Credit	Optimize Process Water Use	2
			Credit	Water Metering	1

0	0	0	**Energy and Atmosphere**	**33**
Y			Prereq	Fundamental Commissioning and Verification	Required
Y			Prereq	Minimum Energy Performance	Required
Y			Prereq	Building-Level Energy Metering	Required
Y			Prereq	Fundamental Refrigerant Management	Required
			Credit	Enhanced Commissioning	6
			Credit	Optimize Energy Performance	18
			Credit	Advanced Energy Metering	1
			Credit	Grid Harmonization	2
			Credit	Renewable Energy	5
			Credit	Enhanced Refrigerant Management	1

0	0	0	**Materials and Resources**	**13**
Y			Prereq	Storage and Collection of Recyclables	Required
			Credit	Building Life-Cycle Impact Reduction	5
			Credit	Environmental Product Declarations	2
			Credit	Sourcing of Raw Materials	2
			Credit	Material Ingredients	2
			Credit	Construction and Demolition Waste Management	2

0	0	0	**Indoor Environmental Quality**	**16**
Y			Prereq	Minimum Indoor Air Quality Performance	Required
Y			Prereq	Environmental Tobacco Smoke Control	Required
			Credit	Enhanced Indoor Air Quality Strategies	2
			Credit	Low-Emitting Materials	3
			Credit	Construction Indoor Air Quality Management Plan	1
			Credit	Indoor Air Quality Assessment	2
			Credit	Thermal Comfort	1
			Credit	Interior Lighting	2
			Credit	Daylight	3
			Credit	Quality Views	1
			Credit	Acoustic Performance	1

0	0	0	**Innovation**	**6**
			Credit	Innovation	5
			Credit	LEED-Accredited Professional	1

0	0	0	**Regional Priority**	**4**
			Credit	Regional Priority: Specific Credit	1
			Credit	Regional Priority: Specific Credit	1
			Credit	Regional Priority: Specific Credit	1
			Credit	Regional Priority: Specific Credit	1

0	0	0	**TOTALS**	Possible Points:	**110**

Certified: 40-49 points, **Silver:** 50-59 points, **Gold:** 60-79 points,
Platinum: 80-110 points

The scorecards were referenced from United States Green Building Council (USGBC)

LEED v4.1 BD+C: Retail
Project Checklist

Project Name:
Date:

Y ? N

0	0	0	Credit	Integrative Process	1

Y	?	N		**Location and Transportation**	**16**
0	0	0	Credit	LEED for Neighborhood Development Location	16
			Credit	Sensitive Land Protection	1
			Credit	High-Priority Site and Equitable Development	2
			Credit	Surrounding Density and Diverse Uses	5
			Credit	Access to Quality Transit	5
			Credit	Bicycle Facilities	1
			Credit	Reduced Parking Footprint	1
			Credit	Electric Vehicles	1

Y	?	N		**Sustainable Sites**	**10**
0	0	0	Prereq	Construction Activity Pollution Prevention	Required
Y			Credit	Site Assessment	1
			Credit	Protect or Restore Habitat	2
			Credit	Open Space	1
			Credit	Rainwater Management	3
			Credit	Heat Island Reduction	2
			Credit	Light Pollution Reduction	1

Y	?	N		**Water Efficiency**	**12**
0	0	0	Prereq	Outdoor Water Use Reduction	Required
Y			Prereq	Indoor Water Use Reduction	Required
Y			Prereq	Building-Level Water Metering	Required
Y			Credit	Outdoor Water Use Reduction	2
			Credit	Indoor Water Use Reduction	7
			Credit	Optimize Process Water Use	2
			Credit	Water Metering	1

Y	?	N		**Energy and Atmosphere**	**33**
0	0	0	Prereq	Fundamental Commissioning and Verification	Required
Y			Prereq	Minimum Energy Performance	Required
Y			Prereq	Building-Level Energy Metering	Required
Y			Prereq	Fundamental Refrigerant Management	Required
			Credit	Enhanced Commissioning	6
			Credit	Optimize Energy Performance	18
			Credit	Advanced Energy Metering	1
			Credit	Grid Harmonization	2
			Credit	Renewable Energy	5
			Credit	Enhanced Refrigerant Management	1

Y	?	N		**Materials and Resources**	**13**
0	0	0	Prereq	Storage and Collection of Recyclables	Required
Y			Credit	Building Life-Cycle Impact Reduction	5
			Credit	Environmental Product Declarations	2
			Credit	Sourcing of Raw Materials	2
			Credit	Material Ingredients	2
			Credit	Construction and Demolition Waste Management	2

Y	?	N		**Indoor Environmental Quality**	**15**
0	0	0	Prereq	Minimum Indoor Air Quality Performance	Required
Y			Prereq	Environmental Tobacco Smoke Control	Required
Y			Credit	Enhanced Indoor Air Quality Strategies	2
			Credit	Low-Emitting Materials	3
			Credit	Construction Indoor Air Quality Management Plan	1
			Credit	Indoor Air Quality Assessment	2
			Credit	Thermal Comfort	1
			Credit	Interior Lighting	2
			Credit	Daylight	3
			Credit	Quality Views	1

Y	?	N		**Innovation**	**6**
0	0	0	Credit	Innovation	5
			Credit	LEED-Accredited Professional	1

Y	?	N		**Regional Priority**	**4**
0	0	0	Credit	Regional Priority: Specific Credit	1
			Credit	Regional Priority: Specific Credit	1
			Credit	Regional Priority: Specific Credit	1
			Credit	Regional Priority: Specific Credit	1

Y	?	N		**TOTALS**	**Possible Points:**	**110**
0	0	0				

Certified: 40-49 points, **Silver:** 50-59 points, **Gold:** 60-79 points,
Platinum: 80-110 points

The scorecards were referenced from United States Green Building Council (USGBC)

227

LEED v4.1 BD+C: Schools

Project Checklist

Project Name:
Date:

Y ? N

Credit	Integrative Process				1

Location and Transportation — 15

Credit	LEED for Neighborhood Development Location	15
Credit	Sensitive Land Protection	1
Credit	High-Priority Site and Equitable Development	2
Credit	Surrounding Density and Diverse Uses	5
Credit	Access to Quality Transit	4
Credit	Bicycle Facilities	1
Credit	Reduced Parking Footprint	1
Credit	Electric Vehicles	1

Sustainable Sites — 12

Prereq	Construction Activity Pollution Prevention	Required
Prereq	Environmental Site Assessment	Required
Credit	Site Assessment	1
Credit	Protect or Restore Habitat	2
Credit	Open Space	1
Credit	Rainwater Management	3
Credit	Heat Island Reduction	2
Credit	Light Pollution Reduction	1
Credit	Site Master Plan	1
Credit	Joint Use of Facilities	1

Water Efficiency — 12

Prereq	Outdoor Water Use Reduction	Required
Prereq	Indoor Water Use Reduction	Required
Prereq	Building-Level Water Metering	Required
Credit	Outdoor Water Use Reduction	2
Credit	Indoor Water Use Reduction	7
Credit	Optimize Process Water Use	2
Credit	Water Metering	1

Energy and Atmosphere — 31

Prereq	Fundamental Commissioning and Verification	Required
Prereq	Minimum Energy Performance	Required
Prereq	Building-Level Energy Metering	Required
Prereq	Fundamental Refrigerant Management	Required
Credit	Enhanced Commissioning	6
Credit	Optimize Energy Performance	16
Credit	Advanced Energy Metering	1
Credit	Grid Harmonization	2
Credit	Renewable Energy	5
Credit	Enhanced Refrigerant Management	1

Materials and Resources — 13

Prereq	Storage and Collection of Recyclables	Required
Credit	Building Life-Cycle Impact Reduction	5
Credit	Environmental ProductDeclarations	2
Credit	Sourcing of Raw Materials	2
Credit	Material Ingredients	2
Credit	Construction and Demolition Waste Management	2

Indoor Environmental Quality — 16

Prereq	Minimum Indoor Air Quality Performance	Required
Prereq	Environmental Tobacco Smoke Control	Required
Prereq	Minimum Acoustic Performance	Required
Credit	Enhanced Indoor Air Quality Strategies	2
Credit	Low-Emitting Materials	3
Credit	Construction Indoor Air Quality Management Plan	1
Credit	Indoor Air Quality Assessment	2
Credit	Thermal Comfort	1
Credit	Interior Lighting	2
Credit	Daylight	3
Credit	Quality Views	1
Credit	Acoustic Performance	1

Innovation — 6

Credit	Innovation	5
Credit	LEED-Accredited Professional	1

Regional Priority — 4

Credit	Regional Priority: Specific Credit	1
Credit	Regional Priority: Specific Credit	1
Credit	Regional Priority: Specific Credit	1
Credit	Regional Priority: Specific Credit	1

TOTALS — Possible Points: 110

Certified: 40–49 points, **Silver:** 50–59 points, **Gold:** 60–79 points, **Platinum:** 80–110 points

The scorecards were referenced from United States Green Building Council (USGBC)

LEED v4.1 BD+C: Warehouses and Distribution Centers

Project Checklist

Project Name:

Date:

Y ? N

Y	?	N			
0	0	0	Credit	Integrative Process	1

Y	?	N		**Location and Transportation**	**16**
0	0	0	Credit	LEED for Neighborhood Development Location	16
			Credit	Sensitive Land Protection	1
			Credit	High-Priority Site and Equitable Development	2
			Credit	Surrounding Density and Diverse Uses	5
			Credit	Access to Quality Transit	5
			Credit	Bicycle Facilities	1
			Credit	Reduced Parking Footprint	1
			Credit	Electric Vehicles	1

Y	?	N		**Sustainable Sites**	**10**
Y			Prereq	Construction Activity Pollution Prevention	Required
			Credit	Site Assessment	1
			Credit	Protect or Restore Habitat	2
			Credit	Open Space	1
			Credit	Rainwater Management	3
			Credit	Heat Island Reduction	2
			Credit	Light Pollution Reduction	1

Y	?	N		**Water Efficiency**	**11**
Y			Prereq	Outdoor Water Use Reduction	Required
Y			Prereq	Indoor Water Use Reduction	Required
Y			Prereq	Building-Level Water Metering	Required
			Credit	Outdoor Water Use Reduction	2
			Credit	Indoor Water Use Reduction	6
			Credit	Optimize Process Water Use	2
			Credit	Water Metering	1

Y	?	N		**Energy and Atmosphere**	**33**
Y			Prereq	Fundamental Commissioning and Verification	Required
Y			Prereq	Minimum Energy Performance	Required
Y			Prereq	Building-Level Energy Metering	Required
Y			Prereq	Fundamental Refrigerant Management	Required
			Credit	Enhanced Commissioning	6
			Credit	Optimize Energy Performance	18
			Credit	Advanced Energy Metering	1
			Credit	Grid Harmonization	2
			Credit	Renewable Energy	5
			Credit	Enhanced Refrigerant Management	1

Y	?	N		**Materials and Resources**	**13**
Y			Prereq	Storage and Collection of Recyclables	Required
			Credit	Building Life-Cycle Impact Reduction	5
			Credit	Environmental Product Declarations	2
			Credit	Sourcing of Raw Materials	2
			Credit	Material Ingredients	2
			Credit	Construction and Demolition Waste Management	2

Y	?	N		**Indoor Environmental Quality**	**16**
Y			Prereq	Minimum Indoor Air Quality Performance	Required
Y			Prereq	Environmental Tobacco Smoke Control	Required
			Credit	Enhanced Indoor Air Quality Strategies	2
			Credit	Low-Emitting Materials	3
			Credit	Construction Indoor Air Quality Management Plan	1
			Credit	Indoor Air Quality Assessment	2
			Credit	Thermal Comfort	1
			Credit	Interior Lighting	2
			Credit	Daylight	3
			Credit	Quality Views	1
			Credit	Acoustic Performance	1

Y	?	N		**Innovation**	**6**
0	0	0	Credit	Innovation	5
			Credit	LEED-Accredited Professional	1

Y	?	N		**Regional Priority**	**4**
0	0	0	Credit	Regional Priority: Specific Credit	1
			Credit	Regional Priority: Specific Credit	1
			Credit	Regional Priority: Specific Credit	1
			Credit	Regional Priority: Specific Credit	1

0	0	0	**TOTALS**	**Possible Points:**	**110**

Certified: 40-49 points, **Silver:** 50-59 points, **Gold:** 60-79 points,

Platinum: 80-110 points

The scorecards were referenced from United States Green Building Council (USGBC)

Bibliography

City of Seattle. LEED Projects Analysis. Accessed September 2022. https://www.seattle.gov/documents/Departments/OSE/COS_LEED_project_analysis2000-11.pdf.

Cornwall, W. (2007, December 1). Stormwater's Damage to Puget Sound Huge. *Seattle Times*. Accessed September 2022. https://www.seattletimes.com/seattle-news/environment/stormwater-pollution-in-puget-sound-streams-killing-coho-before-they-can-spawn/.

Energy Future, Think Efficiency. (2008, September). American Physical Society. Accessed March 2021. https://journals.aps.org/prxenergy/.

Green Outlook 2011, Green Trends Driving Growth. (2010). McGraw-Hill Construction. Accessed February 2022. https://www.3rsustainability.com/wp-content/uploads/LEED-for-Financial-Institutions-Presented-by-3R-2016.pdf.

International Green Construction Code (IgCC) (2021). A comprehensive solution for high-performance buildings. Accessed March 2022. https://codes.iccsafe.org/content/IGCC2021P1/preface.

International Organization for Standardization. (2006). *ISO 14040 International Standard, Environmental Management, Life Cycle Assessment, Principles and Framework*. International Organization for Standardization.

U.N. Environment Program. State and Trends of the Environment 1987–2001, Section B, Chapter 5. Accessed November 2020. https://www.researchgate.net/publication/260554162_GLOBAL_ENVIRONMENT_OUTLOOK_4_GEO4_chapter05_Biodiversity.

UN-Water. (n.d.). Water Facts. http://www.unwater.org/statistics/en/. Accessed September 2022.

U.S. Environmental Protection Agency (EPA). Water Trivia Facts (2016, February 23). https://water.epa.gov/learn/kids/drinkingwater/water_trivia_facts.cfm. Accessed September 2022.

U.S. Forest Service. Quick Facts (2006, October). https://www.fs.usda.gov/projects-policies/four-threats/facts/open-space.shtml. Accessed September 2022.

U.S. Green Building Council (USGBC). Green Building Benefits (2016, April). http://www.usgbc.org/articles/green-building-facts. Accessed September 2022.

World Resources Institute: Six big benefits from the U.S. inflation reduction Act (2022, July). Banking on a just and green recovery, lessons from nine cities. Accessed September 2022. https://cff-prod.s3.amazonaws.com/storage/files/PGsUJ2iMMuB7xPa6U8xa270lXnoJpH5QIXaVwwxa.pdf.

Index

Note: Page numbers followed by *f* denote figures and by *t*, tables.

233